Data Science for Business and Decision Making: An Introductory Text for Students and Practitioners

Data Science for Business and Decision Making: An Introductory Text for Students and Practitioners

Seyed Ali Fallahchay

ARCLER
PRESS

www.arclerpress.com

Data Science for Business and Decision Making: An Introductory Text for Students and Practitioners

Seyed Ali Fallahchay

Arcler Press
224 Shoreacres Road
Burlington, ON L7L 2H2
Canada
www.arclerpress.com
Email: orders@arclereducation.com

ISBN: 978-1-77407-621-7 (Hardcover)

ABOUT THE AUTHOR

Dr. Seyed Ali Fallahchay completed his PhD. in Business Management from De La Salle Araneta University, Philippines as well as his Masters in Business Administration in Philippines. He received his bachelor from Islamic Azad University, Iran and became licensed Engineer in 2010.

Dr. Seyed Ali Fallahchay currently a professor in Business Administration at Raffles and Design Institute at Jakarta Indonesia. He is also a professor at University of Mansford, California, US. Prior to his current position in academe, he also teaches in different institution in the Philippines. He also became a Senior Program head in Business Administration as well as Research Director. He also became a professor in graduate programs and teaches several business subjects. And he also develops several course materials for graduate programs such as: Production and Operation Management, Quantitative Methods in Business, Methods of Research and Managerial Economics. He was also engaged in the industry and became a successful entrepreneur. He also authored the following books: The Economics of Innovation and Modern Scientific Communication.

TABLE OF CONTENTS

LIST OF FIGURES

LIST OF ABBREVIATIONS

BA	Business Analytics
BI	Business Intelligence
CA	Cambridge Analytica
CPU	Central Processing Unit
CRM	Customer Relationship Management
CU	Capacity Utilization
DBMS	Database Management Systems
DISC	International Symposium on Distributed Computing
DOM	Document Object Model
EDI	Electronic Data Interchange
EITC	Earned Income Tax Credit
ERP	Enterprise Resource Planning
IC	Integrated Circuit
IoT	Internet of Things
KDD	Knowledge Discovery and Data Mining
KPIs	Key Performance Indicators
MES	Manufacturing Execution Systems
MGI	McKinsey Global Institute
ML	Machine Learning
NAICS	North American Industry Classification System
NLP	Natural Language Processing
OEE	Overall Equipment Effectiveness
OLAP	Online Analytical Processing
PODC	Principles of Distributed Computing
RDBMS	Relational Database Management System
RDF	Resource Description Frameworks

ROI	Return on Investment
SGML	Standard Generalized Markup Language
SIGACT	Special Interest Group on Algorithms and Computation Theory
SQL	Structured Query Language
SSEM	Small-Scale Experimental Machine
TCS	Theoretical Computer Science
VLSI	Very-Large-Scale Integration
XML	Extensible Markup Language

PREFACE

We live in a world where information is of the essence. It is a commodity that is bought and sold on the open market. Given the importance of information, it is important to ensure that the quality of its input and output is sufficiently robust to meet the needs of the community at large. Nowhere is the information more important than in business. Many of the business decisions that are being made today rely on information. Indeed, the quality of information can directly influence the quality of decisions that are made. This book is conceived as an introduction to the science of information. Specifically, the book seeks to explore data science as an emerging discipline within the wider context of an information society. This book tends to focus on the role of information in making decisions. In that sense, the book is a partial retort to those who assume that business acumen is really all about gut instinct or intuition. The scientific method has been widely adopted in various fields because it is precise and organization. Science replaces some of the irrationality that used to be a major component of business decision-making. Of course, intuition has a role to play and some people who do not rely on science can make excellent decisions. The only problem is that those excellent decisions occur as a matter chance rather than as a consequence of a well-considered process.

In writing this book, I am not only addressing undergraduate students who may be interested in data science as a possible career option. The book has a much wider audience which includes the very same executives that are responsible for making decisions which have significant implications to the society, well beyond the immediate corporation they may be taking care of. Where decisions are based on facts, they are more likely to be justifiable and less prone to the kind of arbitrariness that can ruin businesses. I am conscious of the fact that data science can be an intimidating discipline for those that have not been trained in it. Therefore, this book is written in a format that is deliberately accessible. At the same time, the book references theory and practice-based knowledge which can be expanded on by those who are pursuing an undergraduate course. The book is written from a paradigm that does not see the age of information as somehow threatening to our cherished traditions. Rather, this book is conceived from the perspective that the information age is a necessity which can transform our businesses and our decision-making if we take the time to embrace it.

In order to achieve the right balance between technical and practical aspects, this book amalgamates the theory of data theory with its practice in a way that is accessible. There is an emphasis on case studies and practical applications as well as grounding in the principles that underpin this field. Moreover, this book looks for a multidisciplinary application of the conceptual issues that are raised by the book. Hence, there is a section that talks about social enterprise. This is a relatively new field that is included in development economics but has traditionally not been a priority for data science. The reason why such a seemingly "alien" section is included is to demonstrate the fact that data science is a dynamic field that is relevant to different people in different situations. Given the size of the book, it is by no means a definitive treatise of data science in business. However, this book provides important insights that should inspire further reading and research.

CHAPTER 1

INTRODUCTION TO DATA SCIENCE

CONTENTS

This initial chapter in the book introduces the readers to the nature and application of data science. The first section focuses on the development of the scientific method and its processes. In the second section, we consider how knowledge is extracted using algorithms. The third section provides the differences and similarities between structured and unstructured data forms. The fourth section highlights the principles of data mining and their application to the big data phenomenon. The chapter concludes with a section on the usage of hardware and software systems.

1.1. THE SCIENTIFIC METHOD AND PROCESSES

There is a general consensus that the scientific method is preferably to other approaches to research (Abu-Saifan, 2012; Chesbrough, 2005; Hair, 2010). For example, there are very people who advocate for using superstition to make business decisions, but we know that many successful entrepreneurs are actually very superstitious and will rely on their intuition to make important business decisions (Awang et al., 2013; Chiu et al., 2016; Hamari et al., 2015). It there seems to be a fallacy to suggest that all business data is acquired through the scientific method and its processes (Bachman, 2013; Davis et al., 2014; Helmreich, 2000). Nevertheless, the experts in the industry advocate for the scientific process because it is verifiable and its results can be replicated in certain circumstances (Bansal, 2013; Dutse, 2013). Besides, the use of the scientific method allows us to develop and apply general insights that might be useful for multiple business channels (Berker et al., 2006; Ellison, 2004; Hilbert and Lopez, 2011). Right from the 17th century, the scientific method has dominated the received wisdom about collecting information (Boase, 2008; Engelberger, 1982; Holmes, 2005). This dominance is mainly explained by the advantages that the method has over other approaches (Cappellin and Wink, 2009; Evans, 2009; Howells and Wood, 1993). For example, the scientific method involves careful observation which is underpinned by healthy skepticism about received wisdom (Carlson, 1995; Gibson and Brown, 2009; Ifinedo, 2016). This skepticism, in turn, ensures that the researcher is very rigorous in their methodologies in order to avoid bias or producing information that is rendered irrelevant by virtue of its inaccuracies (Carr, 2010; Gilks, 2016; Jansen et al., 2008).

The rise of the scientific method is partly linked to the demerits of other approaches which were based on insupportable cognitive assumptions which led to a misreading of reality (Jibril and Abdullah, 2013; Malathy and Kantha, 2013; Mosher, 2013). The scientific method addressed some of this limitation by forming informed hypotheses which could then be tested in order to ascertain whether the assumptions and expectations are proved or disproved (Kees et al., 2015; McFarlane, 2010; Noughabi and Arghami, 2011). In this way, knowledge was constantly being tested and improved with time (Kim and Jeon, 2013; Menke et al., 2007; Rachuri et al., 2010). That is why the scientific method does not always offer a definitive answer to questions, but rather an answer that is the best on offer given the circumstances and other information that is readily available (Kirchweger et al., 2015; Mieczakowski et al., 2011; Ruben and Lievrouw, n.d.). For example, it could be hypothesized that downsizing during a recession is highly desirable since it reduces the costs that the company has to bear (Kobie, 2015; Miller, 2014; Sakuramoto, 2005). However, such an assertion may not be permanently accurate if the government is offering stimulus packages to businesses based on their workforce numbers that are at risk (Lewis, 1996; Min et al., 2008; Sin, 2016). In this case, an assumption that has been held to be true over time-based on existing empirical knowledge is then modified by new realities (Little, 2002; Sinclaire and Vogus, 2011). The scientific method can therefore not be static but must constantly find ways to continuously test knowledge (Lyytinen et al., 2016; Min et al., 2009; Sobh and Perry, 2006). It is hard work following the scientific method and that is why some businesses end up ignoring the science and instead relying on traditional approaches to decision-making including intuition (Spiekermann et al., 2010; Stone et al., 2015). Figure 1.1 summarizes the processes of the scientific method which have made it such a successful source of knowledge.

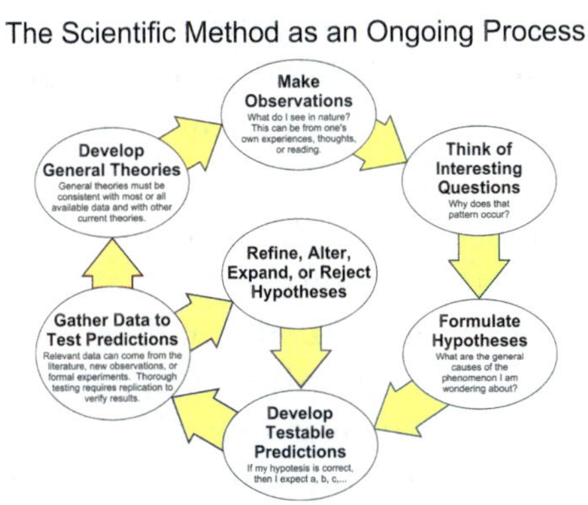

Figure 1.1. The scientific method.

Source: Wikimedia Commons.

When the scientific method is inductive, it will generate new knowledge from careful observations that highlight certain patterns about life (van Nederpelt and Daas, 2012). For example, a small business will after many years of observing the purchase patterns conclude that there are seasons when small purchases are likely to occur whereas, at other times, the consumers are most likely to make bulk purchases (van Deursen et al., 2014). This will then inform the business decision-making process which is based on real data rather than supposition or speculation (Zhang and Chen, 2015). At other times, the scientific method is experimental and reflects the reality that business or organizational life is not always predictable at the first instance (Wallace, 2004). In that sense, the search for information becomes a case of trial and error (Ulloth, 1992). Typically, this is appealing to resilient entrepreneurs who are used to moving ahead even when they are faced with significant setbacks that could potentially become career-modifying (Trottier, 2014). Sometimes the entrepreneur may come up with hypotheses about certain variables that affect their business but the scientific method will need to test those hypotheses on a regular basis in order to establish whether they hold and in what circumstances they are likely to hold (Tarafdar et al., 2014). Eventually, the assumptions and conclusions from the research are refined to the extent that they become functionally useful to the business (Stone et al., 2015). In other words, the scientific method helps to generate business intelligence (BI) with real implications and consequences for all the people that are associated with a particular organization or sector (Spiekermann et al., 2010). Sometimes this knowledge is then disseminated widely in order to influence aggregate behavior or control for otherwise irrational behavior by the main actors within the economy (Sobh and Perry, 2006). At other times, the knowledge is retained and not shared so that it can offer a competitive advantage to the entity that owns it (Sinclaire and Vogus, 2011).

The scientific method has principles that guide its conduct and the way in which its output is to be understood (Sin, 2016). The level of adherence to these principles can vary considerably depending on the purpose for which the scientific method has been deployed (Schute, 2013). For example, a senior researcher at a University will be much more disciplined in their application of the scientific method than a layperson that is merely curious about the behavior of their potential customers (Sakuramoto, 2005). Of course, there are those that suggest that the principles of the scientific method should be strictly adhered to regardless of the situation (Ruben and Lievrouw, n.d.). The businesses that consistently rely on scientific information can achieve the status of scientific enterprises which increases their likelihood of survival, growth, and development (Bachman, 2013). Conversely, those businesses that persist in rely on unscientific methods may occasionally have luck on their side but the odds of them failing are very high (Gibson and Brown, 2009). Over time, many in the industry will develop various models to describe and predict different situations (Ellison, 2004). This becomes the body of practice knowledge that underpins that sector (Helmreich, 2000). The scientific method is also useful to academia when describing and predicting the business world (Gilks, 2016). The advantage of relying on information from academia is that there are specialists whose main objective in their professional careers in scientific research (Howells and Wood, 1993). They, therefore, dedicate many higher-quality resources to the task rather than a business that other operational goals which sometimes supersede the quest for scientific data that may not even be particularly relevant to a given present-day crisis point (McFarlane, 2010).

The downside of relying on information that is produced by academics is that it is often written in ways that may not be practical to the entrepreneur or the people that work for them (Hamari et al., 2015). Academics tend to adopt a very theoretical stance about information and perhaps become practical in terms of gathering empirical evidence (Gibson and Brown, 2009). They are less concerned about producing business solutions, although the trends are changing such that research is often accompanied by so-called "interventions" which are meant to support the business community (Hair, 2010). For the entrepreneur, the main purpose of engaging in the scientific process is to obtain information that will increase their profits and minimize their losses (Howells and Wood, 1993). This type of utilitarian approach can lead to poor research since there is pressure to meet the commercial interests and expectations of the business that is commissioning the research (Lyytinen et al., 2016). The scientific method has greatly benefitted from human curiosity because the search for knowledge is never truly and fully satisfied (Bansal, 2013). Even the most successful businesses know that they have to stay ahead of the competition if they are to have any chance of becoming a major player within their chosen sector (Gilks, 2016). Businesses that were once considered untouchable have collapsed because they were unable to cope with the realities of change (Engelberger, 1982). The dynamic nature of the scientific method, therefore, makes it ideally suited to a business world that is constantly changing and whose competitive spirit demands new solutions for new problems at every turn (Miller, 2014).

Some have argued that one of the problems with modern business is the failure to pay sufficient attention to the theory of entrepreneurship and business (Bansal, 2013). As a consequence, many players end up being instinctive in their decision-making rather than following the scientific method (Boase, 2008). Others argue that a reductivist view of business may not always be appropriate, even where it is counterbalanced by some qualitative research that details the experiences of entrepreneurs from their own perspectives (Hilbert and Lopez, 2011). This point of view bolsters its argument by showing entrepreneurs can succeed without formal education or even significant experience within the sector (Lewis, 1996). Nevertheless, the practice is to incorporate the scientific method in the knowledge search of various enterprises that operate within a modern economy (Kobie, 2015). The amount of credence that these enterprises give to the scientific method can vary from organization to organization and is contextually defined by the challenges that the business is facing at a given moment in time (Schute, 2013). Given the interest in succeeding and the priorities that underpin the modern business sector, many companies have neither the time nor the inclination to engage in extensive longitudinal experiments in order to understand the hypothesis that underpins business theory (Carlson, 1995). Some argue that this is a failing that undercuts the potential that these businesses have to be the best at what they do (Holmes, 2005). In any case, the research and development department in each organization should be tasked with identifying, evaluating, and disseminating scientific information that might be useful to the major decision-makers within the organization (Noughabi and Arghami, 2011).

That then leaves the question of what role conjecture and gut instinct actually plays in business decision-making (Boase, 2008). There is no denying that there are entrepreneurs who are very successful and claim to acquire this success through following their gut instinct. Science may not give credence to their claims, but their success is undeniable (Carlson, 1995). The role of science is therefore to provide an alternative that is predictable and fairly objective (Gibson and Brown, 2009). The resultant information can then be applied to different businesses within different sectors. Ideally, every business should follow the basic steps of the scientific when obtaining information about the operation of their business as follows:

- **Part 1 – Formulation of a Question:** The question can refer to the explanation of a specific observation, as in "Why is the sky blue?" but can also be open-ended, as in "How can I design a drug to cure this particular disease?" This stage frequently involves finding and evaluating evidence from previous experiments (Bansal, 2013), personal scientific observations or assertions (Evans, 2009), as well as the work of other scientists (Menke et al., 2007). If the answer is already known, a different question that builds on the evidence can be posed (Holmes, 2005). When applying the scientific method to research, determining a good question can be very difficult and it will affect the outcome of the investigation (Sobh and Perry, 2006).

- **Part 2 – Hypothesis:** It is a conjecture, based on knowledge obtained while formulating the question that may explain any given behavior (Bachman, 2013; Hair, 2010). The hypothesis might be very specific; for example, Einstein's equivalence principle, or Francis Crick's "DNA makes RNA makes protein" or it might be broad; for example, unknown species of life dwell in the unexplored depths of the oceans. A statistical hypothesis is a conjecture about a given statistical population (Stone et al., 2015). For example, the population might be people with a particular disease. The conjecture might be that a new drug will cure the disease in some of those people. Terms commonly associated with statistical hypotheses are the null hypothesis and alternative hypothesis (Menke et al., 2007). A null hypothesis is a conjecture that the statistical hypothesis is false; for example, that the new drug does nothing and that any cure is caused by chance. Researchers normally want to show that the null hypothesis is false. The alternative hypothesis is the desired outcome that the drug does better than chance. A scientific hypothesis must be falsifiable, meaning that one can identify a possible outcome of an experiment that conflicts with predictions deduced from the hypothesis; otherwise, it cannot be meaningfully tested (Wallace, 2004; Zhang and Chen, 2015; van Nederpelt and Daas, 2012).
- **Part 3 – Prediction:** This step involves determining the logical consequences of the hypothesis (Engelberger, 1982). One or more predictions are then selected for further testing (Boase, 2008). The more unlikely that a prediction would be correct simply by coincidence, then the more convincing it would be if the prediction were fulfilled; evidence is also stronger if the answer to the prediction is not already known, due to the effects of hindsight bias (Chiu et al., 2016; Jibril and Abdullah, 2013; Ruben and Lievrouw, n.d.). Ideally, the prediction must also distinguish

the hypothesis from likely alternatives; if two hypotheses make the same prediction, observing the prediction to be correct is not evidence for either one over the other (Bansal, 2013; Min et al., 2008). These statements about the relative strength of evidence can be mathematically derived using Bayes' Theorem (Hamari et al., 2015; Kobie, 2015).

- **Part 4 – Testing:** This is an investigation of whether the real world behaves as predicted by the hypothesis (Ellison, 2004; Gilks, 2016). Scientists (and others) test hypotheses by conducting experiments (Evans, 2009). The purpose of an experiment is to determine whether observations of the real world agree with or conflict with the predictions derived from a hypothesis (Kim and Jeon, 2013). If they agree, confidence in the hypothesis increases; otherwise, it decreases (Hilbert and Lopez, 2011). The agreement does not assure that the hypothesis is true; future experiments may reveal problems (Min et al., 2008; van Deursen et al., 2014). Karl Popper advised scientists to try to falsify hypotheses, i.e., to search for and test those experiments that seem most doubtful. Large numbers of successful confirmations are not convincing if they arise from experiments that avoid risk. Experiments should be designed to minimize possible errors, especially through the use of appropriate scientific controls (Hamari et al., 2015). For example, tests of medical treatments are commonly run as double-blind tests. Test personnel, who might unwittingly reveal to test subjects which samples are the desired test drugs and which are placebos, are kept ignorant of which are which. Such hints can bias the responses of the test subjects. Furthermore, the failure of an experiment does not necessarily mean the hypothesis is false (Bansal, 2013; Lewis, 1996; Zhang and Chen, 2015). Experiments always depend on several hypotheses, e.g., that the test equipment is working properly, and a failure may be a failure of one of the auxiliary hypotheses. Experiments can be conducted in a college lab, on a kitchen table, at CERN's Large Hadron Collider, at the bottom of an ocean, on Mars (using one of the working rovers), and so on. Astronomers do experiments, searching for planets around distant stars. Finally, most individual experiments address highly specific topics for reasons of practicality (Hilbert and Lopez, 2011). As a result, evidence about broader topics is usually accumulated gradually.

- **Part 5 – Analysis:** This involves determining what the results of the experiment show and deciding on the next actions to take (Evans, 2009). The predictions of the hypothesis are compared to those of the null hypothesis, to determine which is better able to explain the data (Miller, 2014). In cases where an experiment is repeated many times, a statistical analysis such as a chi-squared test may be required (Awang et al., 2013). If the evidence has falsified the hypothesis, a new hypothesis is required; if the experiment supports the hypothesis but the evidence is not strong enough for high confidence, other predictions from the hypothesis must be tested (Bansal, 2013). Once a hypothesis is strongly supported by evidence, a new question can be asked to provide further insight on the same topic (Helmreich, 2000). Evidence from other scientists and experience are frequently incorporated at any stage in the process (Abu-Saifan, 2012). Depending on the complexity of the experiment, much iteration may be required to gather sufficient evidence to answer a question with confidence, or to build up many answers to highly specific questions in order to answer a single broader question (van Deursen et al., 2014).

1.2. KNOWLEDGE EXTRACTION USING ALGORITHMS

Algorithms have emerged as one of the convenient and efficient ways of extracting the kind of information that is useful for businesses when they are making decisions (Lyytinen et al., 2016). In that sense, we are talking about knowledge extraction that involves the creation of useful information from data (Lewis, 1996). It can take the form of structured sources such as XML and relational databases, or alternatively, it can take the form of unstructured sources such as images, documents, and pieces of text (Min et al., 2008). The information that is ultimately extracted has to be placed in a format that can be read and interpreted by the machines that will be making use of it (Tarafdar et al., 2014). That information must also be salient, timely, and ultimately relevant to the original instructions (McFarlane, 2010). This information is then presented in a way that allows for inference (Zhang and Chen, 2015). In methodological terms, this is very similar to the approach that is used in the ETL data warehouse and information extraction NLP (Engelberger, 1982). The difference is that in this latest data extraction, there is an emphasis on going beyond structured information or transformation

(Awang et al., 2013). Instead, the data extraction is carried through to the development of relational schema (Evans, 2009). Figure 1.2 is an example of an adapted knowledge extraction process that relies on algorithms.

Figure 1.2. Algorithmic knowledge extraction process.

Source: Adapted from MDPI.

This type of knowledge extraction relies on pre-existing formal knowledge that is recycled or reused in such a way as to serve the business purposes of the commissioning agent (Davis et al., 2014). A case in point is the reusing of data ontologies or even identifiers (Helmreich, 2000). Alternatively, the knowledge extraction may lead to the development of a new schema that is based on primary source data (Hair, 2010). There are many specialist communities that are actively involved in ensuring that the process for this type of knowledge extraction is streamlined and that the highest standards are maintained at all times (McFarlane, 2010). For example, the RDB2RDF W3C group is currently standardizing a language for extraction of resource description frameworks (RDF) from relational databases (Stone et al., 2015). At the most technical level, the development of these algorithms can seem detached from the world of business which tends to focus on questions and answers that are specifically relevant to the transactional nature of commerce (Engelberger, 1982). However, the acquisition of knowledge then allows the enterprise to interpret it and use it according to the business priorities (Kees et al., 2015). For example, we know that Wikipedia started out as a base source of information that was gathered through co-production with members of the public (Bachman, 2013). However, the use of advanced knowledge extraction has meant that this database is a useful resource for businesses that sometimes also utilize it as a marketing tool in order to showcase their services, products, and even capabilities (Rachuri et al., 2010). This was only possible when the Wikipedia data was transformed into structured data which could be mapped onto existing knowledge about a large range of topics (Trottier, 2014). Some of the programs that were used in this process include Freebase and DBpedia (Holmes, 2005).

One of the over-arching objectives in the sub-sector that is known as knowledge extraction is the development of a standardized knowledge representation language (Berker et al., 2006). That led to the introduction of versions such as OWL and RDF (Ellison, 2004). The development process was underpinned by significant research that also referenced the practice knowledge that had been generated by various businesses (Jansen et al., 2008). There is interest in mapping the transformation of relational databases into a language such as RDF (Cappellin and Wink, 2009). Other areas of interest for researchers include knowledge discovery, identity resolution, and ontology learning (Bansal, 2013). These interests can be labeled topics or niches which are investigated to their fullest extent in order to provide a bigger picture that is rooted in the scientific method and the resultant accuracy (Hamari et al., 2015). Researchers have occasionally used the general process which relies on traditional methods which start with information extraction (Stone et al., 2015). This is then followed by transformation and loading with ETL (Bansal, 2013). The original data sources then become structured formats that can be used by businesses (Kim and Jeon, 2013).

There exists a typology that is used to categorize the range of approaches that are used in algorithmic knowledge extraction (Bachman, 2013). The first categorization may be based on the type of source from which the data is being extracted (Engelberger, 1982). These may include relational databases, text, CSV files, and XML files. The second type of categorization may be based on the nature of exposition (Hair, 2010). This specifically refers to the various ways in which the extracted knowledge is then made explicit so that it is usable for business purposes (Miller, 2014). The ways may include a semantic database or ontology file for example (Helmreich, 2000). Thirdly, the process may be categorized using the reference to the ways in which the information may be queried (Hamari et al., 2015). This is an important element of algorithmic knowledge extraction because it is through the query that the information can be discovered and deployed accordingly (Min et al., 2008). It is possible to categorize algorithmic knowledge extraction using the level and type of synchronization that is ultimately involved (Lyytinen et al., 2016). For example, the process can be executed once in order to produce a dump file (Lyytinen et al., 2016). At other times, you may opt to use a synchronization that is linked to the source (Helmreich, 2000). This is where you can work out whether the synchronization is static or dynamic (Hilbert and Lopez, 2011). Furthermore, you may want to identify whether the changes to the result are written back such that it becomes a bi-directional relationship (Howells and Wood, 1993). Other classifications focus on the reuse of vocabularies during algorithmic knowledge extraction (Engelberger, 1982). The use of an advanced electronic tool or a program in order to recycle the vocabularies that is already in place (Mosher, 2013). Indeed, there are some programs that are able to do this automatically once the initial rules have been clarified during the coding process (Min et al., 2008). Consideration may be given to the level of atomization and whether it has implications for the categorization of the algorithmic knowledge extraction (Boase, 2008). Some of the systems are automatic, semi-automatic, manual or GUI. Finally, the categorization may be based on whether or not the algorithmic knowledge extraction requires domain ontology (Miller, 2014). For example, there are some designs which have a pre-existing ontology which is a prerequisite for proper mapping (Menke et al., 2007). Alternatively, you may have to come up with a new schema which is learned from the source. This is what is known as ontology learning (Zhang and Chen, 2015).

We can look at a technical example in order to understand the background to the algorithmic knowledge extraction. Let us start off with an assumption that the various elements of the extraction will use entity linking. The packages that come into play include: DBpedia Spotlight, OpenCalais, Dandelion dataTXT, the Zemanta API, Extractiv, and PoolParty Extractor. The task is to analyze free text via named-entity recognition and then disambiguates candidates via name resolution. The links are found and then passed on to the DBpedia knowledge repository (Dandelion dataTXT demo or DBpedia Spotlight web demo or PoolParty Extractor Demo). If for example, the chairman of the Federal Reserve in the USA announces an interest rate cut; there will be implications for many businesses that may not be able to visit all the relevant pages for analysis. Using algorithmic knowledge extraction, it is possible to bring together the various strands of information which are then presented as a unit of knowledge that can be used to help with decision making about borrowing and lending. There will be many technical conversion of information before the final product is presented to the decision-maker (Bansal, 2013). That decision-maker has to be educated enough and experienced enough to pick out the information that is relevant to them (Mieczakowski et al., 2011). The use of algorithmic knowledge extraction is particularly important in the information age where BI is a significant competitive advantage (Ellison, 2004). Those organizations that are able to acquire the best information are going to always be a lot more competitive than their counterparts that are relying on conjecture or even second-hand information (Kees et al., 2015).

The challenge for most businesses is being able to reconcile the technical aspects of algorithmic knowledge extraction and the day-to-day information that is necessary for them to make appropriate business decisions (Evans, 2009). At this stage, it is important to recognize the fact that in order to make a business decision that is informed; you do not have to be a mathematical genius or a wizard of database management (Gilks, 2016). Part of the process of administering the information is ensuring that managers and other decision-makers are presented with information that is of interest to them and which they can rely on (Howells and Wood, 1993). At the same time, measures must be put in place to ensure that decision-makers take an interest in the output that comes from algorithmic knowledge extraction (Howells and Wood, 1993). Those that are more technologically savvy can take one extra step by trying to understand the technical process (Howells and Wood, 1993). This will give them insights that they can use in order to make inferences about the information that comes to them (Min et al., 2008). Moreover, having this information is not a guarantee of accuracy and completeness (Kobie, 2015). The best organizations will look for supplementary information that they can use to complement their efforts elsewhere (Helmreich, 2000). The end result is a fairly comprehensive picture of the knowledge and information that they need in order to run their businesses well (Malathy and Kantha, 2013). There are also problems of cost and absorption for the smaller enterprises which may not have the luxury of hiring an expert to assist them (Miller, 2014). In those cases, it is important to consider the merits and demerits of pooled resources (Lyytinen et al., 2016). The algorithmic knowledge extraction produces knowledge that is already out there and is actually applicable to many organizations (Sin, 2016). It, therefore, makes sense for these organizations to occasionally work together in order to be able to access this vital information (Helmreich, 2000).

Despite its technical undertones, the process of algorithmic knowledge extraction plays a critical role in the overall business decision process for any organization (van Deursen et al., 2014). It is one of the starting points for gathering up to date information that is relevant before making decisions (Chiu et al., 2016). If the process is not right, then there is a high likelihood that the outcomes of the algorithmic knowledge extraction will also not be right (Kirchweger et al., 2015). The impact of gathering inaccurate or irrelevant data is not just about being ill-equipped to make decisions, but also making decisions that have far-reaching consequences for the industry as a whole and the business in particular (Menke et al., 2007). The democratization of the internet has meant that virtually anybody can come up with information in the virtue world and distribute it widely (Cappellin and Wink, 2009). Businesses that take everything that is on the internet for the gospel truth are bound to make mistakes because they are misreading the very nature of the platform from which they are obtaining information (Davis et al., 2014). The process of algorithmic knowledge extraction has a number of limitations, the principal of which is the inability to replicate the sense of judgment that a live human being might have (Miller, 2014). Therefore, algorithmic knowledge extraction merely aggregates what is available depending on the inclusion and exclusion criteria that have been written into the respective aggregates (Trottier, 2014). If the primary information is somehow inappropriate for a given decision, the process of algorithmic knowledge extraction will not be able to correct that problem (Noughabi and Arghami, 2011). It might highlight some inconsistencies or provide a very wide view of the entire data list (Sinclaire and Vogus, 2011). However, that in no way means that it will be able to correct the issues or even make human judgments about the kind of information that is required (Engelberger, 1982). That still remains a role for human beings, a role that they are tasked with continuing despite the presence of algorithmic knowledge extraction within the arsenal of business resources (Menke et al., 2007).

1.3. INSIGHTS INTO STRUCTURED AND UNSTRUCTURED DATA

When businesses are confronted with data, it can be either structured or unstructured (Awang et al., 2013). This characteristic of the data will have a direct impact on how they interpret and use that data (Gibson and Brown, 2009). Unstructured data is sometimes considered to be on the raw side of data because it does not have particular patterns and categorizations that might be useful when making decisions (Howells and Wood, 1993). Structured data will already have been broken down into its respective subtypes which are clearly distinguishable from each other (Menke et al., 2007). The fact that structured data has clear patterns makes it easier to search for and find (Menke et al., 2007). Unstructured data then comprises all other types of data which is neither easily broken down, nor easily searchable (Noughabi and Arghami, 2011). Unstructured data may be presented in a number of formats, some of which are incompatible with each other (Min et al., 2008). They could include videos, audio files and even social media posts (Kees et al., 2015). At a high level, it may be possible to have a single post that contains many different unstructured data sets (Min et al., 2009). The challenge then is to break them down into something that is comprehensible and ultimately usable (Ruben and Lievrouw, n.d.). Figure 1.3 is an example of structured and unstructured data about the 2016 Presidential Election in the USA.

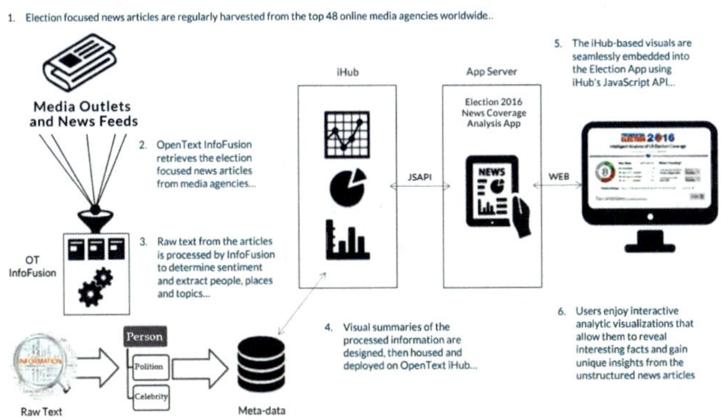

Figure 1.3. Example of structured and unstructured data.

Source: BCAFOT.

There is virtually no tolerance for a real conflict between structured and unstructured data (Boase, 2008). Consumers may select either one of them but this decision is not determined by the way in which the data has been structured or unstructured (Gibson and Brown, 2009). Instead, consumers will mainly focus on the application that uses the data (Zhang and Chen, 2015). The more convenient and helpful applications will most likely attract more customers than those applications that are not user-friendly (Awang et al., 2013). Typically, relational databases tend to make use of structured data while the other applications are dominated by unstructured data (Gibson and Brown, 2009). The big data companies engage in a process of curating and distributing this data to specific outlets and consumers depending on their preferences (Malathy and Kantha, 2013). The more established organizations or businesses with complex informational networks may require big data solutions (Chiu et al., 2016). This means that they have to find a service provider that is experienced in this field and has the capacity to meet their needs (Gibson and Brown, 2009). Although the structured and unstructured data sets are not in conflict in terms of their presentation, the tension comes when choosing an appropriate analytical strategy (Jibril and Abdullah, 2013). For example, most analysts find it easier to deal with structured data and yet a large proportion of data is actually unstructured (Little, 2002). That means that the entity that commissions the analysts will have to pay more to deal with unstructured data (Min et al., 2008). These price differentials are justified by the fact that structured data analytics is considered to be a more mature process with set procedures that are well-established, researched, and tested (McFarlane, 2010). Moreover, a lot of technology that is on the market today is geared towards dealing with structured data (Engelberger, 1982). The implication is that unstructured data analytics is bound to remain a nascent industry for the foreseeable future and that is will mean that it requires a lot of research and development investment (Gibson and Brown, 2009). Moreover, the technologies that are currently being used to deal with unstructured data are not mature (Helmreich, 2000).

Businesses have to make a decision as to what type of analytical technology and human resources are appropriate for their needs (Bachman, 2013). Others are trying to engage in a process of triangulation which involves aggregating both structured and unstructured data in order to improve the BI that is available to the organization (Bansal, 2013). There have been a number of traditional assumptions that underpin the way in which data is handled (Chiu et al., 2016). For example, it is assumed that structured data will be deposited in relational databases or RDBMS. Fields store length-delineated data phone numbers, Social Security numbers, or ZIP codes (Gibson and Brown, 2009). Even text strings of variable length like names are contained in records, making it a simple matter to search (Miller, 2014). Data may be human- or machine-generated as long as the data is created within an RDBMS structure (Malathy and Kantha, 2013). This format is eminently searchable both with human-generated queries and via algorithms using the type of data and field names, such as alphabetical or numeric, currency or date (Kirchweger et al., 2015). Common relational database applications with structured data include airline reservation systems, inventory control, sales transactions, and ATM activity (Kim and Jeon, 2013). Structured query language (SQL) enables queries on this type of structured data within relational databases (Mosher, 2013). Some relational databases do store or point to unstructured data such as customer relationship management (CRM) applications (Jansen et al., 2008). The integration can be awkward at best since memo fields do not loan themselves to traditional database queries (Ifinedo, 2016). Nevertheless, the vast majority of the CRM data is structured (Ellison, 2004).

The definitions of unstructured data in existing literature tend to categorize it as everything else that is not structured data (Chesbrough,

2005). Although this type of data might have an internal structure, it is not organized like would be the case in structured data (Chiu et al., 2016). There are no pre-defined schemas or data models that can easily be turned into queries or references (Ifinedo, 2016). This kind of data may be both textual and non-textual (Gilks, 2016). Other types of unstructured data are generated by human beings while there are varieties that are generated by machines (Noughabi and Arghami, 2011). Indeed, unstructured data can be stored in non-relational database systems such as NoSQL (Mieczakowski et al., 2011). Typical human-generated unstructured data includes:

1. **Text Files:** Word processing, spreadsheets, presentations, email, logs (Awang et al., 2013; Gilks, 2016; Hair, 2010).
2. **Email:** It has some internal structure thanks to its metadata, and we sometimes refer to it as semi-structured (Hair, 2010; Howells and Wood, 1993; Miller, 2014). However, its message field is unstructured and traditional analytics tools cannot parse it (Hair, 2010; Mosher, 2013; Wallace, 2004).
3. **Social Media:** Data from Facebook, Twitter, LinkedIn (Davis et al., 2014; Jansen et al., 2008; Sakuramoto, 2005).
4. **Website:** YouTube, Instagram, photo sharing sites (Bansal, 2013; Helmreich, 2000; Min et al., 2008).
5. **Mobile Data:** Text messages, locations (Abu-Saifan, 2012; Berker et al., 2006; Carr, 2010).
6. **Communications:** Chat, IM, phone recordings, collaboration software (Awang et al., 2013; Boase, 2008; Chesbrough, 2005).
7. **`Media:** MP3, digital photos, audio, and video files (Bachman, 2013; Cappellin and Wink, 2009; Chiu et al., 2016).
8. **Business Applications:** MS Office documents, productivity applications (Bansal, 2013; Carlson, 1995; Davis et al., 2014).

Typical machine-generated unstructured data includes:

1. **Satellite Imagery:** Weather data, landforms, military movements (Dutse, 2013; Ellison, 2004; Wallace, 2004).
2. **Scientific Data:** Oil and gas exploration, space exploration, seismic imagery, atmospheric data (Ellison, 2004; Little, 2002; Zhang and Chen, 2015).
3. **Digital Surveillance:** Surveillance photos and video (Gibson and Brown, 2009; Ulloth, 1992; van Deursen et al., 2014).

4. **Sensor Data:** Traffic, weather, oceanographic sensors (Awang et al., 2013; Hilbert and Lopez, 2011; Schute, 2013).

When big data analysis wants to be inclusive and comprehensive, it must necessarily include both structured and unstructured data sets (Davis et al., 2014). Besides the obvious difference between storing in a relational database and storing outside of one, the biggest difference is the ease of analyzing structured data vs. unstructured data (Holmes, 2005). Mature analytics tools exist for structured data, but analytics tools for mining unstructured data are nascent and developing (Kees et al., 2015). Users can run simple content searches across textual unstructured data (Min et al., 2008). But its lack of orderly internal structure defeats the purpose of traditional data mining tools, and the enterprise gets little value from potentially valuable data sources like rich media, network or weblogs, customer interactions, and social media data (Spiekermann et al., 2010). Even though unstructured data analytics tools are in the marketplace, no one vendor or toolset are clear winners (McFarlane, 2010). And many customers are reluctant to invest in analytics tools with uncertain development roadmaps (Jansen et al., 2008). On top of this, there is simply much more unstructured data than structured (Menke et al., 2007). Unstructured data makes up 80% and more of enterprise data, and is growing at the rate of 55% and 65% per year (Trottier, 2014). And without the tools to analyze this massive data, organizations are leaving vast amounts of valuable data on the BI table (Wallace, 2004).

Structured data is traditionally easier for Big Data applications to digest, yet today's data analytics solutions are making great strides in this area (Kim and Jeon, 2013). Semi-structured data maintains internal tags and markings that identify separate data elements, which enables information grouping and hierarchies (McFarlane, 2010). Both documents and databases can be semi-structured (Jibril and Abdullah, 2013). This type of data only represents about 5–10% of the structured/semi-structured/unstructured data pie, but has critical business usage cases. Email is a very common example of a semi-structured data type (Miller, 2014). Although more advanced analysis tools are necessary for thread tracking, near-dedupe, and concept searching; email's native metadata enables classification and keyword searching without any additional tools (Lewis, 1996). Email is a huge use case, but most semi-structured development centers on easing data transport issues (Miller, 2014). Sharing sensor data is a growing use case, as are Web-based data sharing and transport: electronic data interchange (EDI), many social media platforms, document markup languages, and NoSQL databases (Noughabi and Arghami, 2011). There are some examples of programs that are used to deal with structured and unstructured data (Dutse, 2013). Businesses must take the time to study the provision available on the market so that they can make purchasing decisions based on the latest intelligence and by assessing the relative merits of all possible programs for their business needs at the time (Lyytinen et al., 2016).

Markup language XML is a semi-structured document language (Holmes, 2005). XML is a set of document encoding rules that define a human- and machine-readable format. Its value is that its tag-driven structure is highly flexible, and coders can adapt it to universalize data structure, storage, and transport on the Web (Jibril and Abdullah, 2013). Open standard JSON (JavaScript Object Notation) JSON is another semi-structured data interchange format (Hair, 2010). Java is implicit in the name but other C-like programming languages recognize it (McFarlane, 2010). Its structure consists of name/value pairs (or object, hash table, etc.) and an ordered value list (or array, sequence, list). Since the structure is interchangeable among languages, JSON excels at transmitting data between web applications and servers (Little, 2002). NoSQL Semi-structured data is also an important element of many NoSQL ("not only SQL") databases (Min et al., 2009). NoSQL databases differ from relational databases because they do not separate the organization (schema) from the data (Hilbert and Lopez, 2011). This makes NoSQL a better choice to store information that does not easily fit into the record and table format, such as text with varying lengths (Gilks, 2016). It also allows for easier data exchange between databases (Ifinedo, 2016). Some newer NoSQL databases like MongoDB and Couchbase also incorporate semi-structured documents by natively storing them in the JSON format (Ifinedo, 2016). In big data environments, NoSQL does not require admins to separate operational and analytics databases into separate deployments (Helmreich, 2000). NoSQL is the operational database and hosts native analytics tools for BI (Min et al., 2008). In Hadoop environments, NoSQL databases ingest and manage incoming data and serve up analytic results (Menke et al., 2007).

These databases are common in big data infrastructure and real-time Web applications like LinkedIn (Little, 2002). On LinkedIn, hundreds of millions of business users freely share job titles, locations, skills, and more; and LinkedIn captures the massive data in a semi-structured format (McFarlane, 2010). When job-seeking users create a search, LinkedIn matches the query to its massive semi-structured data stores, cross-references data to hiring trends, and shares the resulting recommendations with job seekers (Little, 2002). The same process operates with sales and marketing queries in premium LinkedIn services like Salesforce (Malathy and Kantha, 2013). Amazon also bases its reader recommendations on semi-structured databases (Miller, 2014). New tools are available to analyze unstructured data, particularly given specific use case parameters (Sakuramoto, 2005). Most of these tools are based on machine learning (ML) (Ulloth, 1992). Structured data analytics can use ML as well, but the massive volume and many different types of unstructured data require it (Min et al., 2009). A few years ago, analysts using keywords and key phrases could search unstructured data and get a decent idea of what the data involved. eDiscovery was (and is) a prime example of this approach (Kim and Jeon, 2013). However, unstructured data has grown so dramatically that users need to employ analytics that not only works at compute speeds, but also automatically learn from their activity and user decisions (Evans, 2009). Natural language processing (NLP), pattern sensing and classification, and text-mining algorithms are all common examples, as document relevance analytics, sentiment analysis, and filter-driven Web harvesting (Mosher, 2013).

There are many benefits that are associated with utilizing unstructured data analytics with machine-learning intelligence (Ellison, 2004). For example, this program can enable the organization to monitor its internal and external communications for purposes of ensuring that they are in compliance with existing laws and regulations (Mieczakowski et al., 2011). Since the law does not give credence to ignorance, it is imperative for businesses to ensure that they are doing the right things just in case the government decides to engage in a compliance check (McFarlane, 2010). Besides, the money that might have been paid in terms of fines and the resultant loss of reputation means that the use of these networks is justified even if it means hiring new staff or retraining the ones that are already operating in the organization (Miller, 2014). The costs of non-compliance with the regulatory framework may include the potential loss of business opportunities, litigation from third parties, and statutory fines (Ellison, 2004). Businesses have a responsibility to engage in sophisticated data analysis techniques that can recognize pertinent patterns as well as operational monitoring frameworks such as email threading analysis (Min et al., 2009). Large organizations with diverse staffing profiles will be particularly concerned about the potential for inadvertent non-compliance by agents and staff members which lead to vicarious responsibility for the organization (Min et al., 2008). Volkswagen is an example of an organization that was heavily fined through non-compliance following the failure to monitor suspicious messages on chats and email.

The use of this information is not only reactive and preventative, but it can also be developmental where the organization actively seeks to use BI in order to strengthen its position in the market (Berker et al., 2006). A case in point is where companies deliberately track high-volume customer discourses and interactions in order to understand where the trends are pointing (Ellison, 2004). Typically, this happens on social media where memes and other popular formats are given credence depending on the number of people that they can attract (Jibril and Abdullah, 2013). Through the use of sentiment and text analysis, it is possible to understand why some particular topics receive positive or even negative feedback (Bansal, 2013). This is critical information that can help to develop marketing campaigns that are specifically cognizant of these aspects of consumer behavior (Ifinedo, 2016). During those discourses or conversations, it may even be possible to identify a possible threat and neutralize it becomes it starts harming the prospects of the organization (Sinclaire and Vogus, 2011). This type of analysis is a lot more sophisticated than a simple keyword search (Dutse, 2013). The most basic searches will only reveal surface statistics such as the number of times a particular term appears in the queries or listings (Helmreich, 2000). However, at a more advanced level, the analyst will want to know whether the response to that term was positive, neutral, or negative (Kim and Jeon, 2013). They might also want to describe the nature of the discourse surrounding that term including whether or not the participants were conversing with each other (Miller, 2014).

Eventually, there will be a construct of the entire tone of the conversation which will be invaluable for the marketer when they wish to target that particular term (Wallace, 2004). Contextualization is of the essence because a negative reaction in one part of the arena might elicit an entirely different response in another (Mieczakowski et al., 2011). The combination of detail and sophistication is what comprises the improved marketing intelligence that will distinguish the excellent companies from those that are nothing more than mediocre (Hamari et al., 2015). Using ML analytics will help to deal with significant quantities of documents in a relatively short space of time so that management information can be passed on to the decision-makers (McFarlane, 2010). The view that is required in order to make the best decisions includes both a micro and macro outlook (Jansen et al., 2008). Those companies that are even more committed may consider the meso-level and anything in between in order to gain an understanding of the subtleties of consumer behavior (Miller, 2014).

1.4. DATA MINING AND BIG DATA

One of the frequent areas of contention in existing literature concerning data science is the distinction between big data and data mining per se (Carr, 2010). The two terms are related but they are also distinct (see Figure 1.4). It is important for decision-makers to understand the differences so that they know how to deal with each set of data process in a way that is going to be most useful to the organization (Hamari et al., 2015).

The classic definition of big data references the significant volumes of information (structured, semi-structured, and unstructured) which is made available in a large range of arenas (Ellison, 2004). In this case, the term volume of data refers to the size or amount of the output that is coming through (Hamari et al., 2015). Big data is often measured in terms of quintillion figures (Engelberger, 1982). The next aspect to consider is the variety of data that is available.

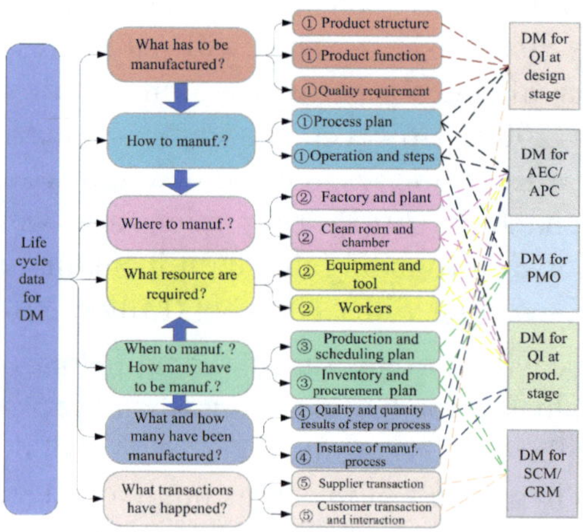

Figure 1.4. Data mining and big data.

Source: MDPI.

There are many sources of data and their output is rarely identical (Kirchweger et al., 2015). For example, a web server log is going to give you a different set of information categories than a social media post; yet both of them may be important for decision-making (Gilks, 2016). Big data involves an element of velocity where the speed of processing and disseminating data is increasing exponentially (Lewis, 1996). Veracity is a more recent concern about big data because of the democratization of the internet (Mieczakowski et al., 2011). Anyone can start disseminating data, but it is another matter to consider whether that data is, actually true (Miller, 2014).

In the age of "fake news," the consequences of not checking veracity can be catastrophic (Mieczakowski et al., 2011). Finally, it is anticipated that big data must add value to the decision-making process (Abu-Saifan, 2012).

It is important to remember that big data is not always a burden to business (Ellison, 2004). Indeed, with the right analysis; this data can be used to improve decision-making and the quality of the implementation process (Gilks, 2016). The most effective businesses will use big data in their strategy moves in order to maximize their competitive advantages and minimize their vulnerabilities (Lyytinen et al., 2016). Because of the popularity of the term, there has been some confusion about the level at which data becomes big data (Berker et al., 2006). Existing literature often emphasizes the sheer volume of the data, more so than its complexity or diversity (Helmreich, 2000). For example, anything that is larger than 1Tb is known as big data (Little, 2002). Moreover, these calculations are sometimes based on the informed estimates of the predicted per capital data (McFarlane, 2010). The latest estimates are that by 2020, per capital data will be 5200 Gbs (Bansal, 2013). Of course, that average does not account for data inequalities with some parts of the world using very little of data while others are overwhelmed with increasing data (Jibril and Abdullah, 2013). We know that on average there are about 50 million tweets sent out in a single day whereas Walmart has to process about 1 million customer transactions every 60 minutes (Awang et al., 2013).

The importance of big data is not so much in terms of its scale as it is in terms of how well it is utilized by various businesses and other entities (Bachman, 2013). Smart decision-making is, therefore, the true performance measure of big data (Dutse, 2013). Such an approach belies those who constantly bemoan the data deluge because it shows that the onus is on them to be selective and to use what they select wisely (Engelberger, 1982). Of course, that is a different matter from the case of a private citizen that is forced to consumer large amounts of irrelevant data because it has been disseminated for marketing purposes (Little, 2002). In that case, the freedom of choice is curtailed by virtue of the fact that powerful companies have accessed the private contact details of the individual and then bombard them with unsolicited marketing material that can significantly hamper their enjoyment of the internet experience (Miller, 2014). Companies have filters and policy frameworks that can control this type of abuse, but private individuals have fewer resources to deal with the problem (Wallace, 2004). In order to make the best out of big data, an organization has to develop a framework with which to receive and analyze the data (Dutse, 2013). The starting point for this process is what is known as data mining (Sakuramoto, 2005). In this instance, the business is actively seeking knowledge and extracting it from the big data that is available on the market (Ulloth, 1992). The data mining procedures start with the large and general before moving towards the small and specific (Noughabi and Arghami, 2011). All other factors being constant, the more specific information is more relevant for business decision-making (Noughabi and Arghami, 2011).

Organizations engage in data mining using a number of schemas some of which include artificial intelligence, ML, and statistics (Engelberger, 1982). The information that is subsequently extracted is used to develop what are known as knowledge discovery databases (Hilbert and Lopez, 2011). Such procedures can be undertaken by private companies or government agencies depending on the circumstances and need (Hamari et al., 2015). The information may be shared with trusted partners in order to engage in a process of cross-referencing (Menke et al., 2007). At other times, the business wants to find out more about the people that have been tracked using their own databases (Miller, 2014). In that sense, data mining can become a coproduction exercise which brings together many companies with mutual interests (Wallace, 2004). Some consumers have objected to this type of approach because they feel that it goes against the spirit of data protection and privacy since the information that is gathered may be used for purposes other than the ones that were declared when it was being collected (Holmes, 2005). There are five levels under which data mining takes place (Engelberger, 1982). The first level involves the extraction of data before it is transformed and loaded onto a data warehouse (Malathy and Kantha, 2013). The second stage involves storing and managing the data. The third stage involves providing access to the right entities in the form of communication (Malathy and Kantha, 2013). On the fourth level, the data is analyzed through a range of processes that are favorable (Menke et al., 2007). The fifth level involves the presentation of the data on a user interface so that they can access and work with it (Zhang and Chen, 2015).

Business engages in data mining for a number of reasons that are not always linear in their objectives (Evans, 2009). For example, the business may want to analyze the various patterns and relationships that signify their current business environment (Little, 2002). The extracted information will help them make better decisions since they are able to work out the reasoning behind consumer behavior through a process of inference and deduction (Noughabi and Arghami, 2011). There are other operational functions that rely on data mining including marketing, credit rating and fraud detection (Engelberger, 1982). There are four relationships that can be identified using data mining (Ifinedo, 2016). The first is known as classes and they are used to locate a target for commercial or other activity (Hilbert and Lopez, 2011). The second relationship is that of clusters in which items are grouped according to their logical location (Helmreich, 2000). The third relationship is known as association whereby the analysts try to map relationships between and among the data sets (Mosher, 2013). The fourth relationship is known as sequential patterning where the analyst is able to anticipate and predict behavior based on past and current trends (McFarlane, 2010).

Despite its demonstrated usefulness over the years, data mining is not without its challenges (Engelberger, 1982). Businesses must be prepared to take on these challenges in order to retain their competitive situation (Gibson and Brown, 2009). The first challenge relates to the range of databases from which data is mined. This represents a challenge for a business that has few resources in terms of expertise, money, and even time (Dutse, 2013). The second challenge relates to the various sources of noise and incomplete data which can send out the wrong information to the analyst (Mieczakowski et al., 2011). As already discussed, poor base knowledge is bound to create poor outcomes (McFarlane, 2010). Some companies have a problem of trying to scale data mining algorithms in a manner that is both efficient and effective. This can become an issue of technical expertise or poor technologies (Wallace, 2004). Other organizations are challenged by the notion of having to handle complex data types that are located within a multiplicity of relationships (Gilks, 2016). It can take a lot of time, patience, and expertise to be able to disentangle the data so that it makes sense from the point of view of making business decisions (Lewis, 1996). In the modern era of electronic threats, data mining can also raise important problems with regards to the integrity of the data, its security and the privacy of the people that are involved (Miller, 2014). The failure to properly address these issues can lead to statutory interventions which may end up costing the organization a lot of money in fines and litigation (McFarlane, 2010).

Given the relationships and linkages between big data and data mining, a comparative analysis can sound like a technical and academic exercise (Carr, 2010). Nevertheless, it is important to acknowledge the distinction because both concepts contribute to the wider whole of data management as a prerequisite to business decision-making (Gibson and Brown, 2009).

Certainly, the people that are in the business of data analysis should know the finer details of the two elements (McFarlane, 2010). Whereas data mining focuses on the details, big data is more concerned about relationships between data (Kim and Jeon, 2013). Data mining, therefore, provides a close-up view of the environment whereas big data is all about the bigger picture (Lyytinen et al., 2016). Data mining expresses the "what" whereas big data express the "why" when confronted with various data classifications (Malathy and Kantha, 2013). Whereas data mining can handle both small and big data sets, big data is almost exclusively at the top end of the volume scales (Bansal, 2013). Data mining is a technique for analyzing data whereas big data is a broader concept that is not so precise in its definition (Lewis, 1996). Data mining can be used to deal with a combination of structured data and databases that are either dimensional or relational (Mieczakowski et al., 2011). By way of contrast, big data deals with structured, semi-structured, and unstructured data (Noughabi and Arghami, 2011). Data mining relies on statistical analysis in order to engage in discovery of small-scale business factors and their prediction (Sakuramoto, 2005). On the other hand, big data handles the macro business factors and predicts large-scale trends (Trottier, 2014). One thing that is clear is the fact that in order to for the BI loop to be completed, both data mining and big data must be referenced at one point or another (Ruben and Lievrouw, n.d.).

The more sophisticated organizations know the precise point at which they can deploy either data mining or big data (Bachman, 2013). This is a subtle decision-making process that relies on a number of inputs which inform the person that is supposed to be making the decision (Kobie, 2015). The first factor to consider is the needs of the organization in terms of understanding its environment (Min et al., 2009). The second consideration is the relative capabilities of the analytical strategies that are under consideration (McFarlane, 2010). The third consideration is the absorptive capacity of the organization itself in terms of expertise, structures, and infrastructure that can support this type of high-level analysis (Stone et al., 2015). In this instance, data is not only used to select specific options among data sets but also how to use the data once it has been selected (Cappellin and Wink, 2009). That is why some researchers note that data is deeply embedded in business decision-making (Helmreich, 2000). Without it, the rationale for the actions of entrepreneurs is somehow diminished (Lewis, 1996). The use of data to support a particular decision may end up sustaining it because the people that are affected by the decision are clear that this decision was based on scientific facts as opposed to speculation and the so-called gut instinct which has hitherto dominated entrepreneurial decision-making (Min et al., 2008). This is particularly important in large organizations where the entrepreneur relies on individual workers to not only fulfill the entrepreneur's vision, but also to go an extra mile in order to ensure that the organization succeeds in its goals and objectives (Hamari et al., 2015).

1.5. USE OF HARDWARE AND SOFTWARE SYSTEMS

The distinction between hardware and software is one of the fundamental aspects of all computer technologies (Bansal, 2013). From a business point of view, it is important to ensure that any purchases hardware is compatible with similarly selected software (Evans, 2009). Both hardware and software must operate competently in order to support business decision-making (Min et al., 2008). The classic definitions of hardware incorporate any physical device that is used in or with the primary computerized product (Little, 2002). For example, the standard personal computer may have additional hardware such as the screen, keyboard, printer, and scanner (Kees et al., 2015). It may also be linked to other related hardware such as the telephony system (Min et al., 2008). On the other hand, software refers to the code that is used to instruct the hardware to perform certain tasks (Zhang and Chen, 2015). This electronic code is typically stored on the hard drive (Hamari et al., 2015). Hence, a typical organization may have a Microsoft Suite as its software incorporating word processing, spreadsheets, and visual presentation software (McFarlane, 2010). No matter how good the software is, it can be let down by the hardware (Hair, 2010). Similarly, the hardware can be let down by faulty software. For example, a computer virus will render the entire system dysfunctional (Miller, 2014). Something similar will happen if the power supply is not getting to the device (Ulloth, 1992). Figure 1.5 is a very basic representation of the links between hardware, software, and organizational operations.

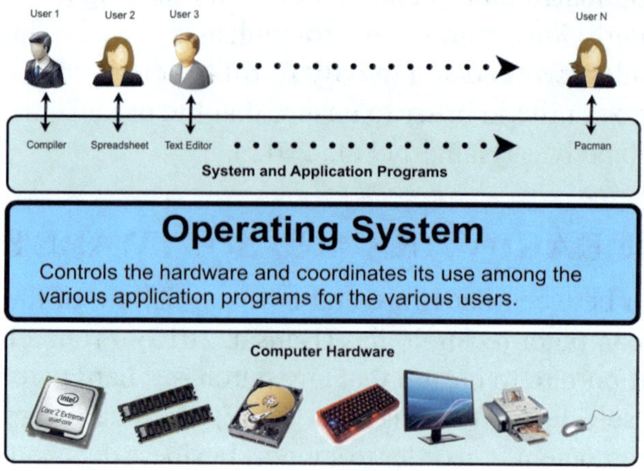

Figure 1.5. Hardware and software in organizations.

Source: Information Technology.

It is important to remember that hardware and software are mutually dependent (Bachman, 2013). No hardware can fully operate without at least one piece of software (Evans, 2009). Similarly, software needs hardware in order to execute the coded commands that it sends out (Kirchweger et al., 2015). For example, in order to prepare a business report, the accounting officers will need Microsoft Word, Excel, and PowerPoint software that will instruct the screen, printer, and keyboard to produce the required report. There are other pieces of hardware and software that may not be easily identifiable within the process but are nonetheless crucial to its success including the central processing unit (CPU), random access memory and the hard drive. Hardware tends to be vulnerable to physical threats such as inclement weather and fire; whereas software is most prone to cyber-attacks such as spyware (Min et al., 2009). Experts argue that the protection of software takes on an even greater level of urgency since a problem can quickly be exported within and outside the organization (Lyytinen et al., 2016). For example, there have been computer viruses that have been able to shut down entire government agencies (Ruben and Lievrouw, n.d.). This is very different from a fire that might destroy specific pieces of hardware in a specific part of the building (van Deursen et al., 2014).

The computer is the base device for many corporations and its hardware is an essential asset that must be catered for (Bachman, 2013). Whereas the CPU acts as the brain for the computer, it is the hardware that actually performs the process of producing output (Bachman, 2013). Hence, for example, the financial ratios are calculated by Microsoft Excel but then start to make sense to the decision-maker once they are sent as an email message or a print-out (Gibson and Brown, 2009). Depending on the needs of the organization, the hardware and software can be reconfigured so that it is adequate to the demands made on it (Jibril and Abdullah, 2013). For example, there are trends where computers will be upgraded by adding new hardware or software components that are deemed to be necessary at the time (McFarlane, 2010). The more advanced organizations can even personalize and patent their internal hardware and software so that it becomes an exclusive competitive advantage when compared to other organizations that are relying on generic products which may not provide the kind of specialization that is required to dominate the market (Gibson and Brown, 2009). The development of modern technologies is ongoing and therefore some of the assumptions that underpinned our understanding of hardware and software have been challenged by new enhancements to computerized systems (Helmreich, 2000). For example, there are situations in which a computer is able to run without software being installed (Little, 2002). However, such situations do not provide the kind of full service that an organization may require in order to make appropriate and effective decisions (Min et al., 2009). A computer without appropriate software will generate an error message and will not be able to produce the kinds of reports that would be required in order to make a business decision (Mieczakowski et al., 2011). A computer needs to have at least some type of operating system that allows both the user and software to interact with the computer hardware (Engelberger, 1982). Installing programs onto the computer in addition to an operating system gives the computer additional capabilities (Bansal, 2013). For example, a word processor is not required, but it allows you to create documents and letters (Noughabi and Arghami, 2011).

Regardless of the type of hardware or software that the firm eventually settles on, matters of security will be important (Bachman, 2013). As already noted, information is an important asset that is used for decision-making (Chiu et al., 2016). That is why some companies have specifically restricted access to the elements that comprise their hardware and software components so that business competitors do not access them (Gibson and Brown, 2009). Access can be used in a number of ways to harm the parent company (Gilks, 2016). For example, a competitor may gain access to a particularly effective computer system and then use it in order to ensure that they have the same competitive advantage as the parent company (Ifinedo, 2016). The acquisition may also involve protecting its own formulas in order to prevent leakages to competitors (Kirchweger et al., 2015). Other competitors are malicious in their use of computerized systems and may eventually plant a virus or another disrupter on the system (Menke et al., 2007). Spyware is of concern because it allows an external entity to control and even monitor all the activities that are taking place within an organization (Dutse, 2013). Copyright rules and regulations may be used in order to ensure that the intellectual rights to the software and hardware are protected (Hilbert and Lopez, 2011). The company has to also regularly upgrade both its hardware and software in order to account for changes in systems as well as new protections against the known threats (Engelberger, 1982). Linked to this, is the need to ensure that the hardware and software that is currently used by the organization is compatible with the other entities that it transacts with (Abu-Saifan, 2012).

Part of the decision-making process for an organization is to select which particular hardware and software computer packages they are going to utilize for a given period of time (Chesbrough, 2005). The market for these products has diversified significantly and the purchaser will be spoilt for choice (Hilbert and Lopez, 2011). However, the fact that there is a lot of choices is not necessarily of assistance since excessive options can lead to poor decision-making (Kirchweger et al., 2015). The starting point for the organization is to set out its minimum requirements and standards (Gilks, 2016). When setting these standards, it is good practice to involve all the departments so that the various facets of the system are addressed (Howells and Wood, 1993). One of the problems that companies face when commissioning hardware and software computer packages is the temptation to use the high visibility and high profile departments as the standard for what is required (Lyytinen et al., 2016). So, they end up purchasing hardware and software computer packages that do not address the entire spectrum of needs that the organization has (Hair, 2010). Eventually, those staff members whose needs were not properly addressed during the commissioning process will lose interest in the entire project and may actually become a liability as they sabotage any packages that have already been bought (Hair, 2010). In order to not lose money, the organization must ensure that everyone is adequately consulted and involved during the commissioning of hardware and software computer packages (Kim and Jeon, 2013).

When preparing a budget for hardware and software computer packages, it is important to consider the maintenance costs (Davis et al., 2014). Some organizations make the mistake of purchasing these products and then not adequately planning for upgrades and breakdowns (Ifinedo, 2016). Ideally, the hardware and software computer packages should be able to operate smoothly throughout the organizational lifecycle (Howells and Wood, 1993). There will be dedicated personnel with the responsibility for ensuring that the hardware and software computer packages are operating as they were intended (Ellison, 2004).

During the maintenance, it is also possible to detect any abuses of the resources such that it can be dealt with at the corporate level (Ifinedo, 2016). A case in point is where workers are making use of free internet in order to engage in social loafing at work (Malathy and Kantha, 2013). When monitoring the activities of the workers, the company must bridge the delicate balance between control and micro-managing (Gibson and Brown, 2009). Some employees respond to a structured format for doing their work and the resultant monitoring or evaluation (Sobh and Perry, 2006). Others work in non-linear and non-predictable ways but end up having to report to people who are used to dealing with highly structured teams (Ellison, 2004). This can cause conflict, particularly within the creative teams which prefer to be given autonomy (Kim and Jeon, 2013).

The employer should carefully weigh the relative risks of control and then provide a form of hardware and software computer packages that can adequately meet the goals of the organization without inconveniencing the people that are supposed to be using these resources (Dutse, 2013). Additional security measures should be put in place to protect the information that is gathered, stored, processed, and distributed by the hardware and software computer packages (Evans, 2009). In fact, this is a requirement under the data protection laws that have been instituted in many jurisdictions (Jibril and Abdullah, 2013). Moreover, it is a commitment to partners and clients to ensure that their information is used according to industry best practices (Lewis, 1996).

CHAPTER 1: SUMMARY

This chapter has introduced the reader to the critical aspects of data science as it relates to decision-making among businesses. The first section showed that the scientific method and its related processes remain the preferred option for businesses that wish to make decisions that are backed by empirical evidence and therefore sustainable in the long run. The second section showed that algorithmic knowledge extraction is an important component of data science because it allows for high-level processing of even the most complicated data sets. The third section highlighted the importance of having distinct strategies for dealing with both structured and unstructured data as it occurs in an organization's environment. The fourth section showed that data mining and big data are all integral, but distinct, processes within the wider concern about data management in modern businesses. The chapter closed by showing that the use of hardware and software computer systems is a complementary process that ensures data quality and data integrity. Overall, the chapter showed that data science is a technical and practical aspect of modern business organizations. The second chapter in this book will be concerned with the use of diverse technologies in data science.

CHAPTER 2

PERIPATETIC AND AMALGAMATED USES OF METHODOLOGIES

CONTENTS

Having explored the basic aspects of data science in the first chapter, the second chapter will consider how peripatetic and amalgamated approaches can be utilized when selecting particular methodologies of data science. The first section will consider how statistical components contribute to data science. In the second section, we will consider the various analytical pathways that can be applied to business data. The third section will introduce machine learning (ML) as a new and important pathway for dealing with business data. This will then lead to a discussion about data-driven science in the fourth section. The chapter will close by summarizing the empirical, theoretical, and computational aspects that underpin data science. The overall objective of this chapter is to enable the practitioner to identify the methodological options for operationalizing the aspects of data science that were introduced in the first chapter.

2.1. STATISTICAL COMPONENTS IN DATA SCIENCE

For many laypeople, data science is synonymous with statistics. Perhaps this association is driven by paranoia about the complexity of statistics (Awang et al., 2013). Indeed, many business decision-makers are reluctant to engage in statistical analysis because of a long-held view that statistics are the confounding variable in their businesses which is used by experts to confuse them and persuade them to spend more money than they need to (Berker et al., 2006). However, statistical analysis has proved to be one of the efficient and simplest ways of turning data science into usable knowledge (Hamari et al., 2015). In order to ensure that decision-makers are not completely repulsed by the notion of statistical analysis, a number of supportive and simplified tools under the umbrella of data analytics have been developed (Hamari et al., 2015). For example, it is now possible for a business to purchase a software package that allows them to input all their transactional data and be presented with clear statistics about their past, current, and predicted performance (Mieczakowski et al., 2011). Data analytics involves analyzing raw data sets in order to select, highlight, and present trends that answer specific research questions that are explicitly set by the client or implied in the brief that is given to the practitioner (Min et al., 2008). Data analytics, therefore, consists of specific goals that determine the field of coverage and the scope of coverage (Little, 2002). There are many technical aspects that underpin data analytics including the tools and techniques that will be deployed for optimum results (Menke et al., 2007). The data analyst must bear in mind the diversity of requests and needs that are set by the consumer (Abu-Saifan, 2012). Some of these needs are not explicitly communicated but must be deciphered by understanding the brief and also studying the environment within which the organization operates (Awang et al., 2013).

It is important to identify the key components for any initiative that involves data analytics. In fact, the data analytics involves combining these different components (Berker et al., 2006). The analysis is supposed to inform their client about the past, present, and future of the organization based on the key components that have been identified (Holmes, 2005). Getting these components might involve liaison work with other members of the staff because they are normally in charge of the units that are supposed to be producing the output that underpins the business data analytics framework (Kobie, 2015). Decision-makers are sometimes looking for a level of clarity which is not entirely possible with the available data set (Boase, 2008). The analyst should be honest and professional enough to describe the merits and limitations of any analytical pathway that they have chosen (McFarlane, 2010). They should also carefully justify any decisions that they make about a specific analytical pathway because they are the kinds of justifications that will eventually determine the sustainability of the selected approach (Jibril and Abdullah, 2013). When all things are constant, the strategies that have yielded the best results are the ones that are most likely to be replicated (Mieczakowski et al., 2011). At the same time, it is important to be flexible because the applicability of a given approach may change with time and other circumstances (Little, 2002). Figure 2.1 highlights the statistical components, their limitations, and their strengths. These are all part of the decision-making calculus which is driven by data.

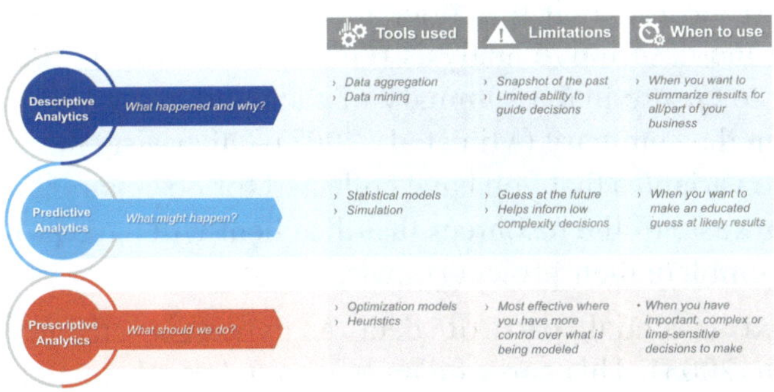

Figure 2.1. Statistical components of data science.

Source: Gurobi.

In most cases, the analyst will begin with a descriptive analysis or descriptive analytics (Boase, 2008). The main purpose of descriptive analytics is to tell the decision-maker about the nature of the information that is being analyzed (van Nederpelt and Daas, 2012). Although some consider descriptive analytics to be unsophisticated, they are actually important when trying to impute and infer wider implications form a given data set (Min et al., 2008). For example, if there are inherent biographical biases in the sample of data; then it can also affect the manner and nature of responses to that data (van Deursen et al., 2014). Indeed, biases at the sampling stage may mean that the premise of the entire report is subsequently called into question (Davis et al., 2014). The most seasoned data analysts ensure that they have all the right information in the right quantities in order to fully support the decision-making process (Zhang and Chen, 2015). Generally, this process begins with descriptive analytics (Bachman, 2013; Ifinedo, 2016). This is the process of describing historical trends in data (Bachman, 2013; Ellison, 2004). Descriptive analytics aims to answer the question "what happened?" (Chesbrough, 2005) This often involves measuring traditional indicators such as return on investment (ROI). The indicators used will be different for each industry (Davis et al., 2014; Holmes, 2005). Descriptive analytics does not make predictions or directly inform decisions (Dutse, 2013; Min et al., 2009). It focuses on summarizing data in a meaningful and descriptive way (Lewis, 1996). There might be a temptation to spend an inordinate amount of time on descriptive analysis but the rule of thumb is the less said, the better (Dutse, 2013). Try to summarize the important facts in a meaningful way so that the decision-maker is not bored by the report before they even begin to read the really critical analytical aspects (Gilks, 2016). Try to always refer back to the research questions or problems that the brief included when doing the descriptive analysis (Gibson and Brown, 2009). It is also a good idea to ensure that a summary of each and every variable collected is included in the summary (Min et al., 2008). Otherwise, the implication is that there are variables that you have collected for no apparent reason which is in effect a waste of the resources that the client will have provided to you in order to complete their project (Lewis, 1996).

The next essential part of data analytics is advanced analytics (Chesbrough, 2005). This part of data science takes advantage of advanced tools to extract data, make predictions, and discover trends (Gilks, 2016). These are important considerations for the decision-maker because they in effect inform them about how the data that has been given to them might be practically useful in their business set up (Gilks, 2016). Given the elevated complexity of this phase, a number of specialized tools may be used in order to analyze the data thoroughly (Ellison, 2004). These tools include classical statistics as well as machine learning (ML) (Lyytinen et al., 2016). ML technologies such as neural networks, natural language processing (NLP), sentiment analysis and more enable advanced analytics (Lewis, 1996). This information provides new insight from data. Advanced analytics addresses "what if?" questions (Menke et al., 2007). It is important to remember that the analyst is looking at the past, present, and future (Kobie, 2015). In the past, they are trying to explain what happened and why it happened. In the present, they are telling the decision-maker about what is happening and why it is happening (Kirchweger et al., 2015). The predictive aspect is the one that looks to the future and will consider the "what if?" questions that the decision-maker may have when they are considering a number of possible options (Sakuramoto, 2005). The availability of ML techniques, massive data sets, and cheap computing power has enabled the use of these techniques in many industries (Holmes, 2005). The collection of big data sets is instrumental in enabling these techniques (Jansen et al., 2008). Big data analytics enables businesses to draw meaningful conclusions from complex and varied data sources, which has made possible by advances in parallel processing and cheap computational power (Mieczakowski et al., 2011). Even where complex information is being communicated, the analyst must always be cognizant of the fact that they may be dealing with decision-makers who are not particularly well-versed in the latest statistical techniques (Chesbrough, 2005). Therefore, the principles of parsimony and simplicity should operate (Little, 2002).

A number of typologies focusing on data analytics have arisen over time (Ellison, 2004). These typologies are associated with the specific functionality that is expected of the data set as well as the methodological limitations that are placed on analysts by their clients (Kirchweger et al., 2015). It is important for the analyst to understand how these typologies are constructed, their uses, advantages, and limitations so that they can advise the client accordingly (Lyytinen et al., 2016). Data analytics is a broad field that has become increasingly complex by virtue of the sophisticated demands that are made with regards to the data and its overall treatment from an analytical perspective (Ellison, 2004). There are four primary types of data analytics: descriptive (Hamari et al., 2015), diagnostic (Kirchweger et al., 2015), predictive (Miller, 2014), and prescriptive analytics (Malathy and Kantha, 2013). Each type has a different goal and a different place in the data analysis process (Malathy and Kantha, 2013). These are also the primary data analytics applications in business (Noughabi and Arghami, 2011).

- **Type I – Descriptive Analytics:** It helps answer questions about what happened (Helmreich, 2000). These techniques summarize large datasets to describe outcomes to stakeholders (Miller, 2014). By developing key performance indicators (KPIs), these strategies can help track successes or failures (McFarlane, 2010). Metrics such as ROI are used in many industries (Little, 2002). Specialized metrics are developed to track performance in specific industries (McFarlane, 2010). This process requires the collection of relevant data (Hamari et al., 2015), processing of the data (Min et al., 2009), data analysis (Kobie, 2015), and data visualization (Little, 2002). This process provides essential insight into past performance (Lyytinen et al., 2016).
- **Type II – Diagnostic Analytics:** It helps answer questions about why things happened (Gibson and Brown, 2009). These techniques supplement more basic descriptive analytics (Gilks, 2016). They take the findings from descriptive analytics and dig deeper to find the cause (Hilbert and Lopez, 2011). The performance indicators are further investigated to discover why they got better or worse (Hamari et al., 2015). This generally occurs in three steps. First, the analyst will identify an emerging problem from the data which is primarily indicated through investigating anomalies (Hamari et al., 2015). These unexpected results may arise out of a particular metric or in a particular market (Kirchweger et al., 2015). The analyst will then investigate the data set that is specifically associated with the anomalies that have been identified (Ellison, 2004). This is almost like a drill-down exercise that seeks to dig underneath the alarming headlines that the initial statistics might indicate (Miller, 2014). The seemingly anomalous data is then analyzed using specialized statistical techniques that will highlight trends and interlink relationships in order to present a mini-narrative of the problem (Mosher, 2013). Diagnostic analytics are very important because it is difficult to solve a problem unless its origin is well understood by the decision-maker (Little, 2002).
- **Type III – Predictive Analytics:** It helps answer questions about what will happen in the future (Bansal, 2013). Businesses that have a good idea of what is going to happen in the future have a very good chance of surviving upcoming challenges and taking advantage of any opportunities that may arise in the future

(Ellison, 2004). This is therefore vital information for decision making in whatever industry is being addressed (Ifinedo, 2016). These techniques use historical data to identify trends and determine if they are likely to recur (Kobie, 2015). Therefore, the prediction is only as good as the information that it relies on (Jibril and Abdullah, 2013). That is why it is imperative to source and keep the most accurate and relevant records for the data analysis (Lewis, 1996). Predictive analytical tools provide valuable insight into what may happen in the future and its techniques include a variety of statistical and ML techniques, such as: neural networks (Kim and Jeon, 2013), decision trees (Holmes, 2005), and regression (McFarlane, 2010).

- **Type IV – Prescriptive Analytics:** It helps answer questions about what should be done (Bachman, 2013). It is really all about providing the client with informed advice so that they can make informed choices (Ellison, 2004). By using insights from predictive analytics, data-driven decisions can be made (Holmes, 2005). This is a very different approach to one where the client comes up with a plan and then asks the analyst to help them deliver that plan (Wallace, 2004). In the case of prescriptive analytics, the plan is actually delivered after reviewing the data and its key messages (Wallace, 2004). This allows businesses to make informed decisions in the face of uncertainty (Hair, 2010). The templates that are eventually used will then assist those that are faced with similar situations in the future which call for decisive action (Kobie, 2015). Prescriptive analytics techniques rely on ML strategies that can find patterns in large datasets (Lyytinen et al., 2016). By analyzing past decisions and events, the likelihood of different outcomes can be estimated (Miller, 2014).

By using these data analytics categories, it is possible for businesses to significantly improve the effectiveness and efficiency of their decision-making at all levels (Awang et al., 2013). Obviously, the decision-making is hierarchical and the information that is provided to the decision-maker will also be determined by their position in the organization's hierarchy (Kees et al., 2015). This calibration of information and its segmentation can be undertaken by the analyst prior to presenting it or by the decision-makers themselves (Lewis, 1996). Alternatively, the organization may set up an internal arbitrator that can decide on who gets what type of information (Noughabi and Arghami, 2011).

These types of data analytics provide the insight that businesses need to make effective and efficient decisions (Bansal, 2013). Used in combination they provide a well-rounded understanding of a company's needs and opportunities (Kim and Jeon, 2013). The role of data analytics is a complex and multidimensional one that exists at the intersection of business decisions, statistical knowledge, and information technology (Malathy and Kantha, 2013). Through the combination of these various categories of competencies and resources, it is possible to provide accurate advice to businesses so that they are able to maximize their chances of survival and success (Jansen et al., 2008). The data analyst is there employed to discover patterns in data and make sense of them before communicating their findings to the decision-makers (McFarlane, 2010). Ultimately, the aim of the data analyst is to ensure that businesses can improve their performance in areas of effectiveness, efficacy, and efficiency (Mieczakowski et al., 2011). The data analysis pipeline is the typical pathway for understanding the information within the environment and then organizing it according to the principles of the profession (Mieczakowski et al., 2011). A successful analyst should be able to work with data in a myriad of ways that reflect the complexity of the environment that they are trying to deconstruct (Sin, 2016). Four processes underpin the work of the data analyst and they can also be used as performance indicators (Gilks, 2016). The first process is that of data mining which is in effect a process of research and discovery (Helmreich, 2000). The second process is that of data management which essentially means organizing and cleaning data so that it is ready for the next process (Lewis, 1996). The statistical analysis allows the analyst to identify, classify, and reconstruct the trends within the data (Gilks, 2016). The final process is that of data presentation in which a report is provided to the client in such a way as to optimize their ability to make good decisions (Mieczakowski et al., 2011). A failure in any or all of the processes will mean that the client has not been served well by their analyst (Min et al., 2008).

Data mining is an essential process for many data analytics tasks (Davis et al., 2014). This involves extracting data from unstructured data sources (Jansen et al., 2008). These may include written text, large complex databases, or raw sensor data (Lewis, 1996). The key steps in this process are to extract, transform, and load data which is often called ETL (Hamari et al., 2015). These steps convert raw data into a useful and manageable format (Helmreich, 2000). This prepares data for storage and analysis (Howells and Wood, 1993). Data mining is generally the most time-intensive step in the data analysis pipeline (Mieczakowski et al., 2011). That is why some companies skip this step, thinking that it is not that important given the fact that its output is in a raw format that is not particularly suitable for sophisticated analysis (Ellison, 2004). However, such an approach would be a mistake since the quality of the final data depends a lot on the quality of the inputs (Hilbert and Lopez, 2011). The company that is commissioning the services of an analyst should request evidence to ensure that the highest standards within the sector are being followed (Malathy and Kantha, 2013). It is not just about finding good sources of data, but also ensuring that once the data is collected; it is treated with the highest professional standards (Jibril and Abdullah, 2013).

Data management or data warehousing is another key aspect of a data analyst's job (Awang et al., 2013). This is the part of the job that must consider the possibilities of presenting the findings of the data mining to the client in a format that is acceptable and helpful to them (Davis et al., 2014). Extra care must be taken to ensure that a user-friendly approach is adopted, particularly with regards to the ways in which the consumer and their representatives are able to integrate the resultant data into their decision-making processes (Kim and Jeon, 2013). Data warehousing involves designing and implementing databases that allow easy access to the results of data mining (Little, 2002). These databases are regularly tested in order to ensure that they are still providing the best support to the decision-maker (Miller, 2014). The analyst will design an initial presentation and then take through a gradual testing process until the optimum configuration has been achieved or approved by the client (Lewis, 1996). This step generally involves creating and managing SQL databases (Bachman, 2013). Non-relational and NoSQL databases are becoming more common as well. Although there are basic principles that should be followed, the nature of data management today allows for some flexibility in terms of changing the rules particularly when those changes are meant to allow the system to function better (Miller, 2014). During the development process, the input of the analyst or someone with a similar professional interest should be included (Kobie, 2015). Likewise, the laypeople that are going to utilize the data should also be involved so that they can highlight certain areas that might need improvement (Sakuramoto, 2005).

Statistical analysis is the heart of data analytics (Hamari et al., 2015). In fact, some organizations give it so much credence that they end up neglecting some of the other aspects of the data handling protocols (Evans, 2009). It is through statistical analysis that predictive frameworks are developed and then eventually shared with the client in order to support their decision-making process (Jibril and Abdullah, 2013). This is how the insights are created from data. Both statistics and ML techniques are used to analyze data (Min et al., 2008). Big data is used to create statistical models that reveal trends in data (Engelberger, 1982). These models can then be applied to new data to make predictions and inform decision making (Awang et al., 2013). Statistical programming languages such as R or Python (with pandas) are essential to this process (Boase, 2008). In addition, open-source libraries and packages such as TensorFlow enable advanced analysis (Helmreich, 2000). Once again, the decision to select a particular software package will depend largely on the needs and experiences of the client (Berker et al., 2006). In some instances, the company will have its own bespoke approach to data analysis and will select tools that meet this criterion (Ellison, 2004). Certainly, there are many products on the market which could be used for this type of work but some organizations have such specific needs that they would prefer to create a bespoke in-house package (Spiekermann et al., 2010). Data analysis is a branch of information technology that is constantly changing and there is no reason to believe that the market will not eventually find something suitable for even the most demanding clients (Menke et al., 2007).

The final step in most data analytics processes is data presentation (Hilbert and Lopez, 2011). This is an essential step that allows insights to be shared with stakeholders (Ifinedo, 2016). Data visualization is often the most important tool in data presentation (Kirchweger et al., 2015). Compelling visualizations are essential to tell the story in the data which may help executives and managers understand the importance of these insights (Miller, 2014). Given the fact that some managers are already biased against the use of business analytics (BA), the strategic use of presentation can be used to win them over (Gilks, 2016). It is important not to overlay the report with technical details which may not be understood by the decision-maker (Noughabi and Arghami, 2011). The first choice for many people when they are presented with incomprehensible information is to simply ignore it and yet this information could be vital for the decision that is being taken (Ulloth, 1992). At the same time, the analyst must be wary of the temptation to over-simply complex issues that need to be taken very seriously in order to make appropriate business decisions (Gibson and Brown, 2009). For example, there is a very good reason for configuring financial ratios in the way that they are (Helmreich, 2000). Changing them does a disservice to the industry as well as the decision-maker (Howells and Wood, 1993). Financial analysts must also be wary of the practice of giving the client the news that they want to hear in the vain hope that the "feel good" effect will sustain the contract and inspire the business to even better achievements (Kim and Jeon, 2013). The role of the analyst is not to manipulate the decision-maker, but rather to provide them with enough information to make their decision (Malathy and Kantha, 2013).

Despite some of the misgivings about the way in which data analytics are used, it remains one of the key business principles and support frameworks (Evans, 2009). The applications of data analytics are very broad (Holmes, 2005). Typically, firms have reported that when done well, data analytics can significantly improve efficiency across the board (Malathy and Kantha, 2013). This holds true regardless of whether the data analytics are done at the unit level of whether they adopt a more generic strategic stance (Min et al., 2009). The competitiveness of the commercial world has implications for the ability of businesses to survive and data analytics can become a competitive advantage when used accurately (Awang et al., 2013). This is particularly important when the organization is dealing with uncertainty or other stresses within the industry (Gilks, 2016). One of the earliest adopters is the financial sector (Min et al., 2009). For example, even before the financial crises of 2008–2011 were taking place; some businesses that had invested in detailed predictive data analytics were aware that a crash was about to happen. The enterprising organizations were able to insulate themselves from some of the worst effects of the crises. Data analytics has a huge role in the banking and finance industries, used to predict market trends and assess risk (Lyytinen et al., 2016). Credit scores are an example of data analytics that affects everyone (Jibril and Abdullah, 2013). A credit score has become such an important aspect that some people consider it to be on part with a passport (Lewis, 1996). At other times, the analytics are accused of being unfair in their prediction because they are not an exact science (Cappellin and Wink, 2009). That is what happens with a credit score which imputes intention to pay based on past behavior without accounting for the notion that people change over time due to experience and better financial circumstances (McFarlane, 2010). These scores use many data points to determine lending risk. Data analytics is also used to detect and prevent fraud to improve efficiency and reduce the risk for financial institutions (Malathy and Kantha, 2013).

The use of data analytics goes beyond maximizing profits and ROI (Ellison, 2004). Data analytics can provide critical information for healthcare in the form of health informatics (Davis et al., 2014), crime prevention (Kirchweger et al., 2015), and environmental protection (Miller, 2014). These applications of data analytics use these techniques to improve our world (Evans, 2009). Therefore, it could be argued that data analytics have a purpose beyond their technical capabilities (Ifinedo, 2016). Though statistics and data analysis have always been used in scientific research, advanced analytic techniques and big data allow for many new insights (Kobie, 2015). Through wider applications, organizations can expand the ways in which data analytics are used in the real world (Holmes, 2005). These techniques can find trends in complex systems (Mosher, 2013). As the world becomes more complex, the application of data analytics might be one of the ways in which the opaquer aspects are illuminated and made accessible to the business leaders that need them in order to make decisions (Jibril and Abdullah, 2013). A case in point is how researchers are currently using ML to protect wildlife. The use of data analytics in healthcare is already widespread (Lewis, 1996). Predicting patient outcomes, efficiently allocating funding and improving diagnostic techniques are just a few examples of how data analytics is revolutionizing healthcare. The pharmaceutical industry is also being revolutionized by ML (Little, 2002). Drug discovery is a complex task with many variables which may be hard to turn into business decisions, particularly if the decision-maker has limited experience of working in business. ML can greatly improve drug discovery. Pharmaceutical companies also use data analytics to understand the market for drugs and predict their sales.

Other uses of data analytics are more in line with the traditional usage of information technologies (Hamari et al., 2015). The internet of things (IoT) is a field that is exploding alongside ML. These devices provide a great opportunity for data analytics (Gibson and Brown, 2009). IoT devices often contain many sensors that collect meaningful data points for their operation. Devices like the Nest thermostat track movement and temperature to regulate heating and cooling. Smart devices like this can use data to learn from and predict your behavior. This will provide advanced home automation that can adapt to the way human beings live (Holmes, 2005). Despite all the applications of business data analytics, it is important to understand their limitations in a wider world that makes a lot of demand on businesses to perform beyond expectations (Evans, 2009). A case in point is where businesses are so reliant on business data analytics that they forget some of the other competitive advantages that they may have such as exceptional customer care (Kees et al., 2015). Others are so wedded to the assumed perfection of business data analytics that they fail to take a critical stance when assessing the information that has been provided to them, hence losing some business opportunities in the process (McFarlane, 2010). An example is where businesses make assumptions about people who have poor credit rating that they fail to recognize the potential that the subprime market has (Malathy and Kantha, 2013). Eventually, the people that are being systematically excluded will start to seek alternatives which will, in turn, supplant the original business that has been relying on business data analytics (Min et al., 2008).

There is also some public concern about the power that enterprises can have over the public by virtue of all the information that they have (Engelberger, 1982). Consequently, there may be a backlash in the form of people withdrawing from those services that seem to expose them to scrutiny from entities that ought to have no business checking their personal details (McFarlane, 2010). The market for consumer contact details is rising and some of the techniques that are used are not entirely moral or ethical (McFarlane, 2010). Most people have at one point or another dealt with unsolicited and therefore unwanted marketing messages from companies that assume that they have a very good idea of the consumer's tastes and preferences. Members of the public are concerned about appearing in a range of databases which are used for purposes that are well beyond their comprehension (Kirchweger et al., 2015). The internet has made anonymity more difficult than it used to be. News and misinformation travel fast (Lewis, 1996). The data analysts then pick it up and aggregate it into management information which helps to make decisions that have profound effects on society as a whole (Min et al., 2009). The rapid speculative movements of the stock movement have exasperated investors across the globe for this same reason (Rachuri et al., 2010). Some decision-makers have complained about being overwhelmed with all types of data without consideration about their ability to process the information deluge (Engelberger, 1982). In other words, there is a danger that real decision-making is going to be subjugated to the whims of data analysts that are not always well versed in the needs of business and who may have other motives that drive them to abuse the process of data analytics (Kirchweger et al., 2015). These remain questions and answers for every modern entrepreneur (Ruben and Lievrouw, n.d.).

2.2. ANALYTICAL PATHWAYS FOR BUSINESS DATA

The selection of analytical pathways is one of the important aspects of decision-making when handling data (Bachman, 2013). This is based on the assumption that BA involves an iterative and methodical exploration of various data sets before they are subjected to statistical analysis in order to make important inferences that support decision-making (Evans, 2009). Various business data analytical pathways have emerged in existing literature (Jansen et al., 2008). The companies that tend to invest in this process are the ones that rely on decisions that are driven by data (Jansen et al., 2008). In that sense, the business data becomes a valuable corporate asset that can be leveraged when trying out to compete other players within the industry or sector (Malathy and Kantha, 2013). Data quality is crucial when trying to make data-driven decisions (Jibril and Abdullah, 2013). The principle of garbage-in-garbage-out applies herein as far as the quality of the decision-making is a direct result of the quality of the inputs that are used when making that decision (Miller, 2014). It is equally important to recruit, train, and direct skilled analysts that have experience of dealing with the specific research problems that a given corporation may be experiencing (Miller, 2014). Technologies play a role because, in the modern era of big data, it is unreasonable to expect that all the transactional elements will be undertaken manually (Little, 2002). The organization itself must be committed towards using the data that is available in a specific way that helps it to gain insight into its environment, therefore making decisions that are optimum considering the variables that determine success as an outcome (Trottier, 2014). The selection of the right business data analytical pathways can also help to overcome the challenges that are faced by real-time enterprises (see Figure 2.2).

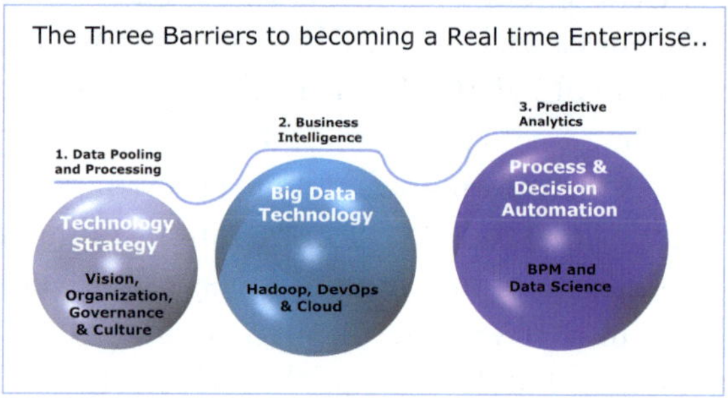

Figure 2.2. Challenges for real-time enterprises.

Source: Vamsi.

The previous chapter already discussed some of the basic typologies of BA including the descriptive analytics that track KPIs in order to paint a picture of the organization's present (Evans, 2009); predictive analytics which focus on trends that give us insights into the likelihoods for the organization based on its current or past state (Kobie, 2015); and prescriptive analytics which recommend interventions based on past performance or the projected future (Miller, 2014). The client must clearly set out the goals of the business analysis before any procedure begins. The analyst will then advise them about the most appropriate methodology given the goals that have been expressed (Bachman, 2013). Having established those two elements, the data acquisition can commence prior to the analytical phases (Gilks, 2016). In data acquisition, there is often an element of extraction which will refer to one or more businesses (Kim and Jeon, 2013). There has to be a protocol for cleaning and integrating the data so that it is now held in a single repository that is accessible to the analyst and possibly the client as well (Jibril and Abdullah, 2013). A case in point is the use of data marts and data warehouses that are used by organizations in order to store that data which is relevant to their business interests (Min et al., 2008). They can then select data sets from it which they will analyze depending on the research problem that has been identified (Sinclaire and Vogus, 2011). It is important that the repository is updated on a regular basis since the data keeps changing with time (Hair, 2010).

The analytical process is not a one-off superficial undertaking (Abu-Saifan, 2012). Rather it is methodical process that has specific steps that must be fulfilled in order to successfully undertake the other future stages (Awang et al., 2013). The initial analysis, for example, will involve a small representative sample of the wider data set (Bachman, 2013). Here the initial trends might emerge or some of the problems in the original data could be identified (Bansal, 2013). There are many analytical tools that might be used at this stage including spreadsheets and advanced statistical functions (Cappellin and Wink, 2009). Other corporations may use complex data mining as well as predictive modeling software (Evans, 2009). Once new relationships and patterns are discovered, the analyst can set even more research questions based on the initial findings (Boase, 2008). This is an iterative process until all the goals that were set out in the brief are fully met (Miller, 2014). The deployment of predictive models involves scoring data records (Kees et al., 2015). These are typically stored and processed in a database (Mieczakowski et al., 2011). The analyst will use the scores to optimize real-time decisions within applications and business processes (Miller, 2014). BA also supports tactical decision-making in response to unforeseen events (Noughabi and Arghami, 2011). There are many instances where for practical and pragmatic reasons, the process of decision-making is automated to support real-time responses (Mieczakowski et al., 2011). In such instances, it would be a major handicap for a business if it could not gain access to up to date information when needed (Kobie, 2015).

Existing discourse sometimes mixes the meaning of BA and BI (Engelberger, 1982). Although the two terms have links and inter-relationships, they are also different from each other in important ways which might affect how they are used in decision-making (Ellison, 2004). The more advanced areas of BA can start to resemble data science, but there is also a distinction between these two terms (Ruben and Lievrouw, n.d.). Even when advanced statistical algorithms are applied to data sets, it doesn't necessarily mean data science is involved (Hair, 2010). That is because true data science involves more custom coding and exploring answers to open-ended questions (Gilks, 2016). Data scientists generally do not set out to solve a specific question, as most business analysts do (Gilks, 2016). Rather, they will explore data using advanced statistical methods and allow the features in the data to guide their analysis (Kees et al., 2015).

There are a host of BA tools that can perform these kinds of functions automatically, requiring few of the special skills involved in data science (Jibril and Abdullah, 2013).

BA refers to the skills, technologies, practices for continuous iterative exploration and investigation of past business performance to gain insight and drive business planning (Lyytinen et al., 2016). BA focuses on developing new insights and understanding of business performance based on data and statistical methods (Ruben and Lievrouw, n.d.). In contrast, BI traditionally focuses on using a consistent set of metrics to both measure past performance and guide business planning, which is also based on data and statistical methods (Little, 2002). BA can contribute to BI when they are specifically geared towards the unique traits of a given industry or an organization within an industry (Holmes, 2005). For example, we might know that there is a drought and that it is affecting crop output. This is known through BA.

However, the information that is gleaned from BA can be turned into BI when it relates to a farmer that is dealing in perishable goods that require minimum rainfall. They will understand that they have a unique position in that particular market which can be exploited by leveraging the hardy nature of their chosen crop. At the same time, they will also be aware of the risks that are involved in keeping perishable goods for long so they will most likely find markets and consumers that are willing to engage in relatively quick purchases.

BA makes extensive use of analytical modeling and numerical analysis including explanatory and predictive modeling and fact-based management to drive decision making (Howells and Wood, 1993). It is therefore closely related to management science (Gibson and Brown, 2009). Analytics may be used as input for human decisions or may drive fully automated decisions (Holmes, 2005). BI is querying, reporting, online analytical processing (OLAP), and "alerts" (Kees et al., 2015). In other words, querying, reporting, OLAP, it is alert tools can answer questions such as what happened, how many, how often, where the problem is, and what actions are needed (Hair, 2010). BA can answer questions like why is this happening, what if these trends continue, what will happen next (predict), and what is the best outcome that can happen (Gibson and Brown, 2009). One of the examples where BA is changing its operation is that of healthcare (Hair, 2010). Traditional healthcare has adopted a curative approach that intervenes when there is a problem (Gibson and Brown, 2009). However, such an approach is expensive because it deals with problems when they have turned into crises (Hamari et al., 2015). BI is then required to ensure that the healthcare industry can support potential clients in efficient manner (Little, 2002). Hospitals and other healthcare units have therefore developed performance indicators that allow them to engage with clients in a much more informed way. For example, once someone is diagnosed as suffering from a lifestyle disease such as TYPE II diabetes, the BI that is available to the hospital will also help to provide additional support to the client who may end up significantly improving their lifestyle as a consequence of their experiences.

When making decisions about a particular analytical pathway, it is important to make reference to certain principles of research which include efficacy, parsimony, beneficence, and simplicity (Kirchweger et al., 2015). Although these are typically used in scientific research for academic purposes, they are also important for business research that involves analyzing extensive data (Little, 2002). Efficacy is a composite term that references both the efficiency and effectiveness of the pathway that has been selected (Chiu et al., 2016). In effectiveness, we are referring to the ability of the pathway to measure what the client wants to be measured after the goals of the organization have been discussed with the analyst (Gilks, 2016). The second element of efficacy relates to efficiency. The pathway must achieve the goals of the organization without expending unnecessary resources (Bachman, 2013). Sometimes analysts insulate themselves from the realities that the client is facing in terms of their budget (Gibson and

Brown, 2009). This can mean making recommendations about analytical pathways that are simply not realistic because they cost more than the utility that they provide to the client (Jibril and Abdullah, 2013). The analyst must consider how useful the information that they are going to provide is going to be when taking into consideration the immediate and future circumstances of the organization (Little, 2002). Sometimes prudence involves cutting out necessary analytical pathways that muddle the waters and could end up confusing the client (Sakuramoto, 2005). In some cases, the very decision to a data analysis approach may be based on a cost-benefit consideration which is undertaken prior to commissioning the project (Berker et al., 2006). Businesses are aware of the benefits of data analysis, but they may still opt-out if they come to the conclusion that the costs involved are not justified by the maximum potential benefits that they can gain from such an approach (Hamari et al., 2015). Moreover, the data analyst will be competing with traditional approaches such as gut instinct which the entrepreneur may be reluctant to abandon (McFarlane, 2010).

The principles of simplicity and parsimony are related because they both focus on options that will achieve the best results for the least effort (Davis et al., 2014). In other words, when applying the principle of parsimony, the analyst should seek the least complicated and resource-consuming approach to achieve the goals that have been set out in the brief (Lyytinen et al., 2016). There are many elaborate and technical approaches in data analysis but this does not mean that they have to all be used in every report (Little, 2002). Although analysts are required to have knowledge of all the approaches and techniques, they have the discretion to recommend simpler approaches if they are bound to achieve the same results (McFarlane, 2010). Once again, this brings some level of efficiency to the data analytics framework (McFarlane, 2010). At the same time, it is important to remember that complexity is not necessarily bad in situations where the client has to understand issues at a more sophisticated level so that they can deploy an appropriate strategy (Chesbrough, 2005). Therefore, it might be that there is an element of homogenization within a particular sector or industry whereby all the main actors use similar protocols because of the need for sophisticated data analytics (Hair, 2010). The data itself may be complicated and therefore requiring a sophisticated approach. Regardless of whether the analyst adopts a simple or complicated analytical framework, they must continuously consult with the client who will direct them as new information comes to light (Min et al., 2009). The decision-making process is not always linear and may be modified by the initial or interim findings of a data analytics process (Miller, 2014). The principles of simplicity and parsimony are not just restricted to the conceptualization of the data analysis or the pathways that are chosen (Engelberger, 1982). They apply throughout; right from the time the project is commissioned to the time that the report is presented (Kees et al., 2015). The report itself must communicate powerfully without losing the interest of the decision-makers who may have very little technical expertise in the field of data analytics (Min et al., 2009).

The analysts should always be willing to play the guidance role, particularly when they realize that their client is making mistakes out of ignorance or a lack of proper exposure (Davis et al., 2014). BA is neutral in their collection of objective data, but they are also biased by the goals that are expressed within the brief that is provided by the client (Dutse, 2013). The moment the client sends instructions to the analyst, they are immediately also biasing their focus onto the issues that are of particular interest to that organization (Hilbert and Lopez, 2011). Nevertheless, the analyst can have a wider lens than the original project scope (Malathy and Kantha, 2013). For example, they may notice a trend that is not included in the original specification but which might in due course turn out to be either strength or a weakness depending on how the organization is able to respond to it (Lewis, 1996). Having a wider lens will provide the decision-maker with a much more comprehensive base for the final options that they come up with (Min et al., 2009). Meanwhile, those reports that are too restrictive may actually end up preventing the client from making the most appropriate decision or making a decision with incomplete information (Mieczakowski et al., 2011). This is a balancing act which must take into consideration the resources that are available and the express wishes of the client that is paying for the data analysis (Jibril and Abdullah, 2013).

Another consideration is the relative benefits of creating bespoke analytical frameworks rather than relying on open sources (Chiu et al., 2016). Some organizations have decided to aggregate and analyze business data for a fee (Kees et al., 2015). Others offer this information for the public without a specific request for payment (Jibril and Abdullah, 2013). They may sustain their business model through advertising or asking for fees for further detailed analysis (Hair, 2010). The problem with free resources is that they may not actually perform the tasks that the client wants them to perform (McFarlane, 2010). The fact that they are open to the wider public also means that they are not exclusive BI and are therefore less likely to offer any significant competitive advantage since everyone can rely on that data to make appropriate decisions (McFarlane, 2010). Hence, despite the low costs that are involved in these data analytics; they are not always adopted by businesses that are looking for bespoke services that address the issues that are specific to their business or industry (Menke et al., 2007). Other businesses also recognize the role that opens resources can play in an industry (Evans, 2009). Therefore, they adopt a hybrid model in which open information is supplemented or complemented by their own bespoke analytics (Malathy and Kantha, 2013). In that way, the business is not ignorant of what is freely available on the market and yet it is also able to distinguish itself from competitors (Menke et al., 2007). The critics of the hybrid model argue that it misses out on the best of both worlds since it does not adopt a focused stance when selecting an analytical strategy (Gibson and Brown, 2009). They also argue that the hybrid model may be an indicator that the business or the client is not really certain about their role in the sector (Little, 2002).

2.3. MACHINE LEARNING (ML) AS A NEW PATHWAY

ML is the scientific study of algorithms and statistical models that computer systems use to perform a specific task without using explicit instructions, relying on patterns and inference instead (Carlson, 1995; Lewis, 1996; Stone et al., 2015). It is seen as a subset of artificial intelligence (Hamari et al., 2015; Miller, 2014; Trottier, 2014). ML algorithms build a mathematical model based on sample data (Gilks, 2016; Kim and Jeon, 2013; Stone et al., 2015). This is what is known as "training data" (Howells and Wood, 1993; Zhang and Chen, 2015; van Deursen et al., 2014). It is used in order to make predictions or decisions without being explicitly programmed to perform the task (McFarlane, 2010). ML algorithms are used in a wide variety of applications, such as email filtering and computer vision, where it is difficult or infeasible to develop a conventional algorithm for effectively performing the task (Ulloth, 1992). The ubiquitous nature of ML today means that no organization can afford to ignore its possibilities, particularly if that organization has ambitions of dominating the market in the future with new products that give it a competitive advantage (Hilbert and Lopez, 2011). The diversity of ML applications also means that a decision has to be made about which ones are most suitable for the needs of an organization or a client (Min et al., 2008). Figure 2.3 highlights the applicability of ML to the education sector in terms of pedagogy, training, and assessment.

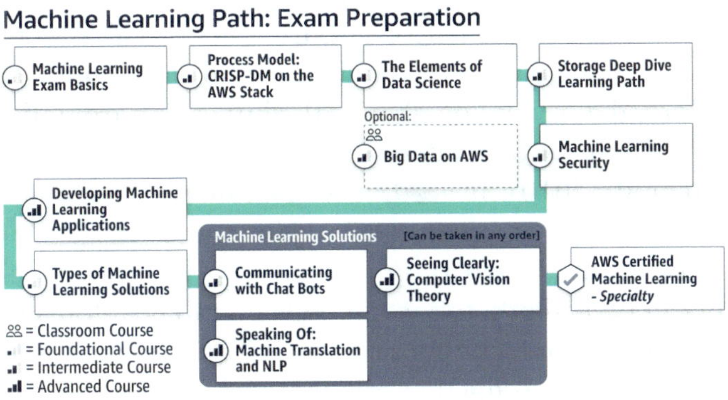

Figure 2.3. The pedagogy of machine learning.

Source: Amazon Web Services.

ML is closely related to computational statistics, which focuses on making predictions using computers (Evans, 2009). The study of mathematical optimization delivers methods (Kirchweger et al., 2015), theory (Min et al., 2008), and application domains to the field of ML (Kobie, 2015). Hence, it is appropriate to consider data mining as a field of study that is in effect a subset of ML. Data mining distinguishes itself from ML by engaging in an exploratory analysis of data using unsupervised learning. That is why some literature refers to ML as predictive analytics based on the notion that it is widely applied for purposes of solving a range of business problems. Arthur Samuel introduced the term "ML" in 1959. However, the term was only fully defined by Tom Mitchell when he summarized the key ingredients of ML as follows:

"A computer program is said to learn from experience E with respect to some class of tasks T and performance measure P if its performance at tasks in T, as measured by P, improves with experience E."

This approach to defining ML focuses on tasks and the operational concerns that proceed from that focus (Engelberger, 1982). Therefore, it can be classified as an operational definition rather than one which is purely conceptual or theoretical (Hamari et al., 2015). There is also some emphasis on the cognitive processes and competencies which are part of what ML seeks to replicate in its operations (Hair, 2010). The central question which Alan Turing attempts to explore is the extent to which a machine can be programmed to think (Malathy and Kantha, 2013). The reality is that no satisfactory answer has been found to that question (Little, 2002). Instead, the focus has moved on whether or not machines are able to replicate the actions of human beings who are known to be thinking entities (Min et al., 2009). Therefore, the discussion about ML shifts away from discussing the humanizing possibilities of modern technology with regards to machines (Spiekermann et al., 2010). Instead, it focuses on the practical capabilities of machines which might approximately replicate the activities of a thinking human being (Kees et al., 2015). For example, it is not so much a case of trying to work out whether a robot can be taught to develop an intellectual interaction with the manufacturing process as it is a case of trying to get the machine to take on the productive role that was once given to the factory worker (Davis et al., 2014).

Turing suggests that there is a possibility of developing a thinking machine which has implications for the role of human beings in the manufacturing process (Gibson and Brown, 2009). In this case, the thinking machine will never quite achieve the intellectual prowess of a functional human being (Gilks, 2016). However, that machine can be given learning tasks which then control its activities which in turn aid the manufacturing process (Helmreich, 2000). A support vector machine is a supervised learning model that divides the data into regions separated by a linear boundary (Howells and Wood, 1993). Here, the linear boundary divides the black circles from the white. ML tasks are classified into several broad categories (Kim and Jeon, 2013). In supervised learning, the algorithm builds a mathematical model from a set of data that contains both the inputs and the desired outputs (Hilbert and Lopez, 2011). For example, if the task were determining whether an image contained a certain object, the training data for a supervised learning algorithm would include images with and without that object (the input), and each image would have a label (the output) designating whether it contained the object (McFarlane, 2010). In special cases, the input may be only partially available, or restricted to special feedback (Ellison, 2004). Semi-supervised learning algorithms develop mathematical models from incomplete training data, where a portion of the sample input doesn't have labels (Jibril and Abdullah, 2013). The approach, in this case, is to try and map the thought processes that go into a specific task before teaching a machine to mirror those processes (Chesbrough, 2005). For example, there will be an intensive study of the actual cognitive and practical processes that go into the assembly line (Helmreich, 2000). Then code will be developed that can trigger or instruct these processes (Hair, 2010). Finally, the process is applied to a machine which can, in theory, perform the tasks that would have been performed by a human being in normal circumstances (Sin, 2016). The downside to those assumptions is the failure to account of the sense of judgment that a human being has and which a machine might never fully replicate (Jansen et al., 2008).

There is an implicit assumption that ML will allow for the performance of tasks that were once the exclusive domains of a human being whilst at the same time avoiding the pitfalls of working with human beings (Gibson and Brown, 2009). For example, human doctors get tired or distracted or biased in their work. Therefore, there is an implicit assumption that a robot might be able to do some aspects of surgery without experiencing the fallibility of human beings. Of course, due to the problems with a sense of judgment and the intangibles of humanity which are hard to replicate by any other entity; there has never been a hospital that is fully staffed by robots without human involvement. Even the mere act of setting up the structures and infrastructure for ML involves the creativity and insight of a human being (Kirchweger et al., 2015). Machine tries to get round some of these challenges through the use of advanced algorithms (Kobie, 2015). For example, it can use classification and regression algorithms. Classification algorithms and regression algorithms are types of supervised learning (Kim and Jeon, 2013). Classification algorithms are used when the outputs are restricted to a limited set of values (Min et al., 2009). For a classification algorithm that filters emails, the input would be an incoming email, and the output would be the name of the folder in which to file the email (Jibril and Abdullah, 2013). For an algorithm that identifies spam emails, the output would be the prediction of either "spam" or "not spam," represented by the Boolean values true and false (Ulloth, 1992). Regression algorithms are named for their continuous outputs, meaning they may have any value within a range (Min et al., 2008). Examples of a continuous value are the temperature, length, or price of an object (Howells and Wood, 1993).

The differences between active and passive ML have also been of interest to researchers in existing literature (Dutse, 2013; Ifinedo, 2016; Rachuri et al., 2010). In unsupervised learning, the algorithm builds a mathematical model from a set of data that contains only inputs and no desired output labels (Lewis, 1996). Unsupervised learning algorithms are used to find structure in the data, like grouping or clustering of data points (McFarlane, 2010). Unsupervised learning can discover patterns in the data, and can group the inputs into categories, as in learning feature (Lyytinen et al., 2016). Dimensionality reduction is the process of reducing the number of "features," or inputs, in a set of data (Jibril and Abdullah, 2013). Active learning algorithms access the desired outputs (training labels) for a limited set of inputs based on a budget, and optimize the choice of inputs for which it will acquire training labels (Dutse, 2013; Jibril and Abdullah, 2013; Zhang and Chen, 2015). When used interactively, these can be presented to a human user for labeling (Hamari et al., 2015). Reinforcement learning algorithms are given feedback in the form of positive or negative reinforcement in a dynamic environment, and are used in autonomous vehicles or in learning to play a game against a human opponent (Berker et al., 2006; Kirchweger et al., 2015).

There may be other specialized algorithms that are selected for a variety of reasons related to the practicality of ML (Evans, 2009). Other specialized algorithms in ML include topic modeling (Lyytinen et al., 2016). This is where the computer program is given a set of natural language documents and finds other documents that cover similar topics (Gibson and Brown, 2009). ML algorithms can be used to find the unobservable probability density function in density estimation problems (Jibril and Abdullah, 2013). Meta-learning algorithms learn their own inductive bias based on previous experience (Jibril and Abdullah, 2013). In developmental robotics, robot learning algorithms generate their own sequences of learning experiences, also known as a curriculum (McFarlane, 2010). This allows them to cumulatively acquire new skills through self-guided exploration and social interaction with humans (Stone et al., 2015). These robots use guidance mechanisms such as active learning (Kim and Jeon, 2013), maturation (Kobie, 2015), motor synergies (van Deursen et al., 2014), and imitation (Schute, 2013).

ML is not an isolated field of study without connection to other fields (Bansal, 2013). Indeed, it is advisable that organizations incorporate it as one of the various data strategies for the modern era (Howells and Wood, 1993). For instance, both data mining and ML have been associated with similar techniques (Gibson and Brown, 2009). The differences between the two terms are more concerned with their focus (Lewis, 1996). In essence, ML focuses on prediction, based on known properties learned from the training data (Dutse, 2013). By way of contrast, data mining focuses on the discovery of previously unknown properties in the data (Hair, 2010). Research has shown that this is the analysis step of knowledge discovery in databases (Gilks, 2016). Data mining uses many ML methods, but with different goals; on the other hand, ML also employs data mining methods as "unsupervised learning" or as a preprocessing step to improve learner accuracy (Jansen et al., 2008). Much of the confusion between these two research communities comes from the basic assumptions they work with (Gibson and Brown, 2009). For example; in ML, performance is usually evaluated with respect to the ability to reproduce known knowledge, while in knowledge discovery and data mining (KDD) the key task is the discovery of previously unknown knowledge (Hilbert and Lopez, 2011). Evaluated with respect to known knowledge, an uninformed (unsupervised) method will easily be outperformed by other supervised methods. However, in a typical KDD task; supervised methods cannot be used due to the unavailability of training data (Mieczakowski et al., 2011).

ML has also been associated with optimization (Carlson, 1995). This occurs because a significant proportion of learning problems come about as a consequence of minimizing a loss function on a given group of samples (Engelberger, 1982). The loss function is actually any discrepancy between the predicted model training outcomes and the empirical evidence of actual outcomes (Kees et al., 2015). A case in point is that of classification when the goal is to assign a label to instances (Jansen et al., 2008). It is assumed that the models which are trained will correctly predict the pre-assigned labels of a set of examples (Kim and Jeon, 2013). However, that is not always the case when the final outcomes are tested (Kees et al., 2015). The difference between the two fields arises from the goal of generalization (Menke et al., 2007). While optimization algorithms can minimize the loss on a training set, ML is concerned with minimizing the loss on unseen samples (Menke et al., 2007). ML and statistics are closely related fields in terms of methods (Dutse, 2013). However, ML is distinct in its principal goal (Ifinedo, 2016). Whereas statistics draw population inferences from a sample, ML finds generalizable predictive patterns (Helmreich, 2000). Michael Jordan suggests that there is a long association between statistics and ML in terms of its methodological principles and theoretical tools (Miller, 2014). That is why some people have suggested that both ML and statistics fall under the larger umbrella that is data science (Sin, 2016).

The theoretical framework that underpins ML is significantly borrowed from two categories (Chiu et al., 2016). The first is known as computational learning theory and the second is known as statistical learning theory (Howells and Wood, 1993). These theories are underpinned by the assumption that the main objective of any learning is to be able to generalize a given experience (Hair, 2010). In this instance, generalization refers to the ability to perform a task in an accurate manner even when dealing with a new and previously unseen set of circumstances (Dutse, 2013). The examples that are used for training purposes are derived from generally known probability distribution patterns that are considered to be a fair approximation or representation of the space of occurrences (Ellison, 2004). The learner then constructs a general model about the space which allows them to produce fairly accurate predictions about new cases (Sobh and Perry, 2006).

The computational analysis of ML algorithms and their performance is a branch of theoretical computer science (TCS) known as computational learning theory (Gilks, 2016). Because training sets are finite and the future is uncertain, learning theory usually does not yield guarantees of the performance of algorithms (Helmreich, 2000; Wallace, 2004; van Deursen et al., 2014). Instead, probabilistic bounds on the performance are quite common (Bansal, 2013; Holmes, 2005; McFarlane, 2010). The bias-variance decomposition is one way to quantify generalization error (Gilks, 2016; Lewis, 1996; Wallace, 2004). For the best performance in the context of generalization, the complexity of the hypothesis should match the complexity of the function underlying the data (Little, 2002). If the hypothesis is less complex than the function, then the model has under-fit the data (Menke et al., 2007). If the complexity of the model is increased in response, then the training error decreases (Sakuramoto, 2005). But if the hypothesis is too complex, then the model is subject to overfitting and generalization will be poorer (Miller, 2014). In addition to performance bounds, learning theorists study the time complexity and feasibility of learning (Abu-Saifan, 2012; Engelberger, 1982). In computational learning theory, computation is considered feasible if it can be done in polynomial time (Bansal, 2013; McFarlane, 2010). There are two kinds of time complexity results. Positive results show that a certain class of functions can be learned in polynomial time (Hilbert and Lopez, 2011). Negative results show that certain classes cannot be learned in polynomial time (Miller, 2014).

2.4. THE USE OF DATA-DRIVEN SCIENCE

Many corporations have come to the conclusion that they have to focus on a management approach that is driven by science (Gibson and Brown, 2009). However, due to the relative ambiguity of the conceptual framework; many firms are not able to articulate this in this mission and vision statements (Engelberger, 1982). In effect, the use of business data-driven science is widespread but not explicitly highlighted in the literature that talks about the operations of various organizations including well-established businesses (Min et al., 2008). Over time, data science has emerged as a multi-disciplinary endeavor in terms of its processes (Evans, 2009), methodologies, techniques, technologies, and systems (Carr, 2010). For example, the algorithms that are used in data science are also applicable in a range of contexts and in a range of disciplines (McFarlane, 2010). The purpose of data science is to ultimately extract useful knowledge from various data sets that may be either structured or unstructured (Jibril and Abdullah, 2013). In reality, this means that data science is a combination of data mining and big data (Kim and Jeon, 2013). It is anticipated that in the modern era, data science will work with the most efficient and advanced algorithms, systems, and technologies in line with its principles of simplicity, parsimony, and effectiveness (Ulloth, 1992). Figure 2.4 demonstrates how the data-driven approach operates in the business world.

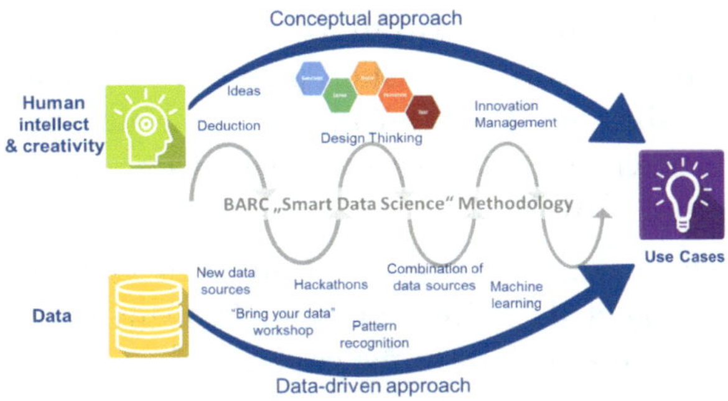

Figure 2.4. The data-driven approach in business.

Source: SAPS Technology Blog.

Some researchers argue that data science is designed to combine the constituent elements of ML, statistical analysis, and related methodologies in order to support decision-making in business (Berker et al., 2006; Gilks, 2016; Stone et al., 2015). In effect, the data analysis helps businesses to understand actual phenomena that is occurring in their environment, but which they may not have properly conceptualized or developed means of monitoring and evaluating (Mieczakowski et al., 2011). Therefore, data science actually facilitates a better fit between a business and its environment through informing the decision-making process (Kobie, 2015). In achieving this objective, data science references a number of theories and deploys different techniques for many fields including computer science, statistics, mathematics, and information science (Hair, 2010). Jim Gray has suggested that data science is a 4th paradigm of science in addition to the other three that include empirical, theoretical, and computational (Howells and Wood, 1993; Lyytinen et al., 2016). Information technology is fueling the rapid changes in science and that is why the term "data deluge" has started to gain credence in both academia and business. In 2015, the American Statistical Association identified database management, statistics, and ML, and distributed and parallel systems as the three emerging foundational professional communities (Lyytinen et al., 2016). These communities then interact with the wider world in order to make data science an integral aspect of our contemporary lifestyle (Hamari et al., 2015). There has been an intersection of data science and popular culture. That is why the 2012 issue of the Harvard Business Review dubbed it the "Sexiest Job of the 21st Century" (Chiu et al., 2016). However, data science was with us for a long time before it became a buzzword. For example, there was once a time when it was referred to as BA, statistics, predictive modeling, and even BI (Malathy and Kantha, 2013). Commentators like Nate Silver argue that at the heart of data, science is statistics and that is really, what it is beyond the hype (Lyytinen et al., 2016).

Data science is an emerging field in academia and professional development (Awang et al., 2013). For example, many universities are now offering the course or variations of the course under different titles (Lewis, 1996). Nevertheless, the key components of the curriculum are still subject to a lot of debate and even disagreement (Helmreich, 2000). One of the problems has been that many data science or big data projects are yet to produce tangible and useful results that change our communities (Ellison, 2004). Some of these problems are not to do with the subject itself, but rather because of the way that it is managed by data wonks that do not have the necessary competences to create systematic knowledge development (Holmes, 2005). Besides, the history of the term is also not particularly linear nor even very clears (Kees et al., 2015). The term "data science" has appeared in various contexts over the past thirty years but did not become an established term until recently (Ifinedo, 2016). In an early usage, it was used as a substitute for computer science by Peter Naur in 1960 who then later introduced the term "datalogy" (Gilks, 2016). In 1974, Naur published the Concise Survey of Computer Methods, which freely used the term data science in its survey of the contemporary data processing methods that are used in a wide range of applications (Malathy and Kantha, 2013). The modern definition of "data science" was first sketched during the second Japanese-French statistics symposium organized at the University of Montpellier II (France) in 1992. The attendees acknowledged the emergence of a new discipline with a specific focus on data from various origins, dimensions, types, and structures (Lewis, 1996). They shaped the contour of this new science-based on established concepts and principles of statistics and data analysis with the extensive use of the increasing power of computer tools (Miller, 2014).

There has always been a fairly strong relationship between data science and statistics, to the extent that the two terms have been used interchangeably (Dutse, 2013). At other times, that close relationship has meant that managers have developed a phobia against data science on account of their concern about the technical aspects of statistics (Jansen et al., 2008). Others suggest that statistics are in effect an instrumentality that enables data science to inform decision-makers (Little, 2002). In that sense, data science has a wider view than statistics although data science cannot exist without statistics (McFarlane, 2010). Business executives have adopted data science as a composite term to describe their data-driven approach to business or the aspects that are skewed towards statistical analysis (Awang et al., 2013). However, there are some academics that argue that there is really no significant distinction between statistics and data science (Kirchweger et al., 2015). Others argue that it is better to be more specific by talking about either data mining or big data (Kim and Jeon, 2013). Gil Press argues that data science is a buzzword without a clear definition and has simply replaced "BA" in contexts such as graduate degree programs. As Nate Silver stated, "Statistics is a branch of science" Data scientist is slightly redundant in some way and people shouldn't berate the term statistician." When it comes to the actual business of running corporations, there is a point of view that data science alone cannot offer complete competitive advantages (Dutse, 2013). Instead, being a data scientist is just one of the four job families that an organization needs in order to leverage big data in its environment (Kobie, 2015). They include data scientists, data analysts, big data engineers, and big data developers (Menke et al., 2007). Therefore, it is not so much a case of ranking the various job roles as ensuring that their participation is included in the mix of employees that a data-driven organization has (Gibson and Brown, 2009).

The classification and categorization of data science is not without its critics (Davis et al., 2014). For example, Irving Wladawsky-Berger sees data science as enjoying the same kind of popular currency that computer science enjoyed when it first came into the scene (Gibson and Brown, 2009). Although currently, data science is multidisciplinary and interdisciplinary in terms of its methodologies and practices; there will come a time when it morphs into a coherent and self-sustaining discipline (Kirchweger et al., 2015). Although computer science is today widely recognized as a critical and self-sustaining discipline, it was once criticized (Dutse, 2013). Others like Vasant Dhar argue that data science is a discipline because it is different from existing analyses across disciplines in as far as it goes beyond merely explaining data sets (Gilks, 2016). In data science, the practitioner seeks a consistent and actionable pattern which they can then use to predictive the behavior of variables (Ellison, 2004). Some of the contributing elements to data science were previously undertaken in other fields and disciplines, but they lacked the kind of systematic theory which was unique to the management of data (Little, 2002). Hence, they could rest in established disciplines such as social science and health sciences without necessarily developing a single unified strand that is data science today (Kobie, 2015).

One of the approaches used to establish data science has to been to critique some of the limited definitions that have hitherto been used in reference to the phenomena (Hair, 2010). David Donoho was particularly concerned about three simplistic definitions that were obscuring the true definition of data science (Gilks, 2016). The first definition tended to equate it to big data (McFarlane, 2010). Specifically, the size of the data or its volume is not an adequate feature to distinguish statistics from science because both of them can deal in small and big data projects depending on the brief that has been provided (Lewis, 1996). The second definition tended to view data science as nothing more than a computational project that sorted and organized big data (Dutse, 2013). The limitation for this definition is the fact that it does not address the peripatetic nature of statistics which can be used by different disciplines (Helmreich, 2000). For example, social statistics have been used for quantitative research projects (Gibson and Brown, 2009). However, there is nothing to say that the statistical models that are used in data science are restricted to none other than that particular discipline (McFarlane, 2010). The third definitional critique focuses on how data science is a heavily applied discipline (Mieczakowski et al., 2011). For example, the vast majority of academic programs that are on offer have been criticized for under preparing their graduates for jobs (Mosher, 2013). This is because the advertisements for these courses are misleading in as far as they suggest that it is statistics training and analytics that are the main and perhaps only components of the course (Jansen et al., 2008). Donoho has been one of the vocal statisticians who have argued that data science needs to be broadened in order to fully account for the professional potential of the discipline (Awang et al., 2013). John Chambers has noted that statisticians to adopt a more inclusive conceptualization of learning from data (Hamari et al., 2015). William Cleveland suggests that the future of data science lies in prioritizing the extraction of data from predictive tools and not in the development or examination of explanatory theories (Kobie, 2015). Ultimately, the wider conceptualization of data science means that it will escape some of its traditional constraints and offer wider functionality for practitioners (Kirchweger et al., 2015).

Moreover, data scientists must account for a future in which coproduction and open science are the norm (Evans, 2009). Although this shifts away from the commercialization of information, it is only with regards to the direct marketing aspects (Menke et al., 2007). The business model is still sustainable by virtue of the peripherals such as advertising and exposure for the sponsors (Kim and Jeon, 2013). Perhaps Wikipedia is the best example of the significant influence that open science can play once the public embraces it (Lyytinen et al., 2016). Even in academia which was once considered to be a largely closed sector, there are still those that advocate for a much wider distribution of literature (McFarlane, 2010). The rationale is that this kind of distribution will open up avenues for engaging with the public and will, therefore, make the respective disciplines that much more appealing than they were during the times that academia was nothing more than an exclusive club of hyper intellectuals (Mosher, 2013). For example, the US National Institute of Health has already announced plans to enhance reproducibility and transparency of research data (Helmreich, 2000). This is in response to research which has shown that laypeople tend to ignore research if it is not presented in formats and avenues that are accessible to them (Gilks, 2016). Whereas academic journals could survive on the basis of a few exclusive subscribers, those elite subscribers were becoming too few to account for the costs of research (Kees et al., 2015). Besides, the market identified a gap and started to produce imitations that served members of the public who wanted to be informed but could not really access some of the more specialized academic journals (Menke et al., 2007). Other big journals are likewise following suit and ensuring that they open up as much as possible without diminishing their high editorial standards (Lyytinen et al., 2016). This way, the future of data science not only exceeds the boundary of statistical theories in scale and methodology, but data science will revolutionize current academia and research paradigms (Mosher, 2013). Donoho argues that the scope and impact of data science will continue to expand enormously in coming decades as scientific data and data about science itself become ubiquitously available (Davis et al., 2014).

2.5. EMPIRICAL, THEORETICAL, AND COMPUTATIONAL UNDERPINNINGS

As already noted, data science is founded on theory, evidence, and practice (Hilbert and Lopez, 2011). It is necessary for practitioners to understand how these elements have come together in order to further the development of the discipline (Trottier, 2014). One such example is that of TCS, a subset of general computer science and mathematics that focuses on more mathematical topics of computing and includes the theory of computation (Hair, 2010; Menke et al., 2007). Although not always initially advertised in undergraduate course units, it can nevertheless be part of the curriculum (McFarlane, 2010). The problem is that theoretical ideas that relate to data science are not always easy to circumscribe (Kirchweger et al., 2015; Tarafdar et al., 2014). According to the ACM's special interest group on algorithms and computation theory (SIGACT), TCS covers a wide variety of topics including algorithms (Chesbrough, 2005), data structures (Helmreich, 2000), computational complexity (Gilks, 2016), parallel, and distributed computation (Holmes, 2005), probabilistic computation (Abu-Saifan, 2012; Dutse, 2013), quantum computation (Lewis, 1996), automata theory (Lewis, 1996), information theory (Lyytinen et al., 2016), cryptography (Ifinedo, 2016), program semantics and verification (Min et al., 2009), ML (Menke et al., 2007), computational biology (Schute, 2013), computational economics (Ruben and Lievrouw, n.d.; Zhang and Chen, 2015), computational geometry (Sakuramoto, 2005), and computational number theory and algebra (Gibson and Brown, 2009; Tarafdar et al., 2014). Work in this field is often distinguished by its emphasis on mathematical technique and rigor (Abu-Saifan, 2012; Bansal, 2013; Helmreich, 2000). Figure 2.5 highlights how data theory, practice, and outcomes can be linked.

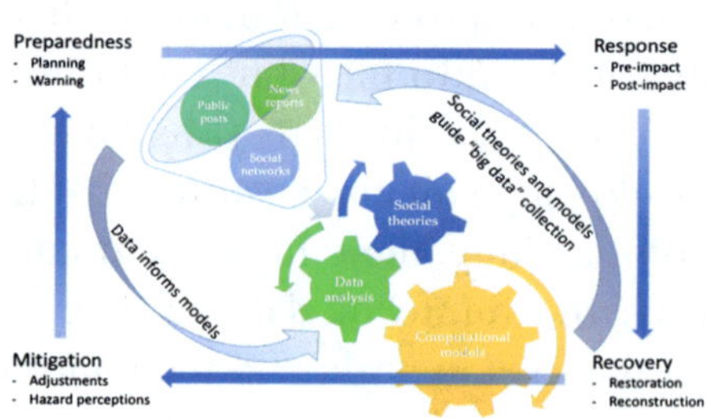

Figure 2.5. Data theory, evidence, and practice.

Source: MDPI.

The fundamentals that underpin data science existed before it was recognized as a separate discipline (Holmes, 2005). For example, Kurt Gödel challenged the existing notions of mathematical proof and logical inference by using an incompleteness theorem (Helmreich, 2000). One of the assumptions of this theorem is that there are statements that can be proved or disproved, a fundamental aspect of the hypotheses in quantitative research (Min et al., 2009). Nevertheless, there were also limitations on what could be and could not be proved; another important distinction that is critical when identifying subjects or variables for scientific study (Hair, 2010). For example, religious dogma is excluded from many scientific studies precisely because it has elements that cannot be proved but must be believed on account of faith (McFarlane, 2010). Eventually, these arguments were converted into discussions about modern logic and computability (Gilks, 2016). Computer science is particularly associated with these discussions. Claude Shannon added an information theory to a mathematical theory of communication in 1948 (Ifinedo, 2016). Around the same time, Donald Hebb unveiled the idea of a mathematical model of learning which could be used to map the workings of the human brain (Zhang and Chen, 2015). There was significant research into the veracity of the claims made concerning early theories about the nature of data (Bansal, 2013; Gilks, 2016; Holmes, 2005). For example, Donald Hebb's hypothesis about a mathematical model of learning which mimics the human brain was increasingly being supported by biological research (Evans, 2009; Malathy and Kantha, 2013; van Deursen et al., 2014). There were, of course, some modifications to the original theory as new information came in (van Nederpelt and Daas, 2012). Consequently, two fields of parallel distributed processing and neural networks starting gaining credence in the academic community (Evans, 2009; Rachuri et al., 2010). Stephen Cook and Leonid Levin proved that there were practically relevant problems that are NP-complete and therefore contributing to the development of the computational complexity theory (Little, 2002). The start of the 20th century saw the development of quantum mechanics (Ifinedo, 2016). The implication was that mathematical operations could be performed on an entire particle wave function (Lewis, 1996). That meant that it was possible to compute functions on multiple states all at the same time (Lewis, 1996). Towards the end of the 20th century, the quantum computer was developed as a specific outcome of this scientific (Sakuramoto, 2005). Such a computer became very popular between 1990 and 2000 (Schute, 2013). Consequently, Peter Shor showed that the same methodological frame could be used to factor large numbers in polynomial time (Mieczakowski et al., 2011). Apparently, such an action could render many public key cryptography systems very insecure (Miller, 2014).

There were other concepts and attributes that had to be defined as a body of theory in order to support the growth of data science (Lewis, 1996). For example, the algorithm received new attention as one of the processing units for computational science (Jibril and Abdullah, 2013). The basic definition of an algorithm is a method that is expressed as a finite list of well-defined instructions which are meant to calculate a function (Lyytinen et al., 2016). An algorithm is an effective method expressed as a finite list of well-defined instructions for calculating a function (Min et al., 2009). Starting from an initial state and initial input, the instructions describe a computation that, when executed, proceeds through a finite number of well-defined successive states, eventually producing "output" and terminating at a final ending state (Tarafdar et al., 2014). The transition from one state to the next is not necessarily deterministic (Sinclaire and Vogus, 2011). That means that there are some algorithms, known as randomized algorithms, which can incorporate random input (Kobie, 2015). Another important concept that has been considered in existing literature is that of data structures (Hair, 2010). A data structure is a particular way of organizing data in a computer so that it can be used efficiently (Jansen et al., 2008). Different kinds of data structures are suited to different kinds of applications, and some are highly specialized to specific tasks (Kobie, 2015). For example, databases use B-tree indexes for small percentages of data retrieval and compilers and databases use dynamic hash tables as lookup tables (Kim and Jeon, 2013). Data structures provide a means to manage large amounts of data efficiently for uses such as large databases and internet indexing services (Menke et al., 2007). Usually, efficient data structures are the key to designing efficient algorithms. Some formal design methods and programming languages emphasize data structures, rather than algorithms, as the key organizing factor in software design (Sobh and Perry, 2006). Storing and retrieving can be carried out on data stored in both the main memory and in secondary memory (Zhang and Chen, 2015).

The computational complexity theory is a branch of the theory of computation that focuses on classifying computational problems according to their inherent difficulty, and relating those classes to each other (Hamari et al., 2015). A computational problem is understood to be a task that is in principle amenable to being solved by a computer, which is equivalent to stating that the problem may be solved by the mechanical application of mathematical steps, such as an algorithm (Hilbert and Lopez, 2011). A problem is regarded as inherently difficult if its solution requires significant resources, whatever the algorithm used (Kees et al., 2015). The theory formalizes this intuition, by introducing mathematical models of computation to study these problems and quantifying the amount of resources needed to solve them, such as time and storage (Kim and Jeon, 2013). Other complexity measures are also used, such as the amount of communication (used in communication complexity), the number of gates in a circuit (used in circuit complexity) and the number of processors (used in parallel computing). One of the roles of computational complexity theory is to determine the practical limits on what computers can and cannot do (Lewis, 1996). Distributed computing studies distributed systems (Kirchweger et al., 2015). A distributed system is a software system in which components located on networked computers communicate and coordinate their actions by passing messages (Lewis, 1996). The components interact with each other in order to achieve a common goal (Mosher, 2013). Three significant characteristics of distributed systems are: concurrency of components, lack of a global clock, and independent failure of components (Menke et al., 2007). Examples of distributed systems vary from SOA-based systems to massively multiplayer online games to peer-to-peer applications (Sinclaire and Vogus, 2011). A computer program that runs in a distributed system is called a distributed program, and distributed programming is the process of writing such programs (Little, 2002). There are many alternatives for the message passing mechanism, including RPC-like connectors and message queues (Mosher, 2013). An important goal and challenge of distributed systems is location transparency (Stone et al., 2015).

Parallel computation is a form of computation in which many calculations are carried out simultaneously (Helmreich, 2000). This process is underpinned by the principle that large problems can often be divided into smaller ones, which are then solved "in parallel" (Kim and Jeon, 2013). There are several different forms of parallel computing. They include the bit-level (Holmes, 2005), instruction-level (Hilbert and Lopez, 2011), data (McFarlane, 2010), and task parallelism (Kim and Jeon, 2013). Parallelism has been employed for many years, mainly in high-performance computing (Holmes, 2005). The interest in parallel computation is due to the physical constraints preventing frequency scaling (Lewis, 1996). As power consumption (and consequently heat generation) by computers has become a concern in recent years (Ruben and Lievrouw, n.d.); parallel computing has become the dominant paradigm in computer architecture, mainly in the form of multi-core processors (Sin, 2016). Parallel computer programs are more difficult to write than sequential ones (McFarlane, 2010). This is because concurrency introduces several new classes of potential software bugs (Hamari et al., 2015). The most common of these bugs are the race conditions (Kim and Jeon, 2013). Communication and synchronization between the different subtasks are typically some of the greatest obstacles to getting good parallel program performance (Holmes, 2005). The maximum possible speed-up of a single program as a result of parallelization is known as Amdahl's law (Little, 2002).

Very-large-scale integration (VLSI) is the process of creating an integrated circuit (IC) by combining thousands of transistors into a single chip (Helmreich, 2000). VLSI began in the 1970s when complex semiconductor and communication technologies were being developed (Holmes, 2005). The microprocessor is a VLSI device (Ifinedo, 2016). Before the introduction of VLSI technology most ICs had a limited set of functions they could perform (Kirchweger et al., 2015). An electronic circuit might consist of a CPU, ROM, RAM, and other glue logic. VLSI allows IC makers to add all of these circuits into one chip (Malathy and Kantha, 2013). Some of these developments are not just restricted to the technical aspects of information technology (Kobie, 2015). Others are associated with disciplines that are outside the word of computational science. One such example is of computational biology (Hilbert and Lopez, 2011). Computational biology involves the development and application of data-analytical and theoretical methods, mathematical modeling, and computational simulation techniques to the study of biological, behavioral, and social systems (Kim and Jeon, 2013). The field is broadly defined and includes foundations in computer science (Ifinedo, 2016), applied mathematics (Kobie, 2015), animation (Kees et al., 2015), statistics (Malathy and Kantha, 2013), biochemistry (Holmes, 2005), chemistry (Jansen et al., 2008), biophysics (Jansen et al., 2008), molecular biology (Hamari et al., 2015), genetics (Jansen et al., 2008), genomics (Schute, 2013), ecology (Miller, 2014), evolution (Spiekermann et al., 2010), anatomy (Sobh and Perry, 2006), neuroscience (Kobie, 2015), and visualization (Wallace, 2004).

It is quite clear from some of the examples that have been presented above that data science is underpinned by complex theories, tools, methods, and techniques (Helmreich, 2000). Despite the fact that data science has been largely associated with the computer geeks and those who are in businesses that specifically sell technology; it seems to be one of the most important professional genres in the future (Kim and Jeon, 2013). Humanity is constantly seeking and finding ways of computerizing activities that were previously undertaken by workers. The era of robots is underpinned by data science as one of the fields that have made the greatest contribution to this category of scientific endeavor (Kobie, 2015). When considering the onset and management of business decisions, the use of data science is not just to facilitate the operations of that organization (Gilks, 2016). Data science can actually produce real products that can be marketed by an organization (Mieczakowski et al., 2011). Indeed, there are many companies that have specialized in this type of product (Mieczakowski et al., 2011). We are only beginning to understanding the value that this type of scientific process can bring to all our activities (Mieczakowski et al., 2011). Even as the focus on data continues, it is always important to place this focus within context (Jibril and Abdullah, 2013). There are many issues that make a successful entrepreneurship venture and data science is one of them (Little, 2002). The decision-makers must find the right balance when selecting the options that can be used in order to arrive at defensible conclusions about their organizations and roles within a given organization (Lyytinen et al., 2016).

CHAPTER 2: SUMMARY

This chapter sought to explain and demonstrate some of the peripatetic and amalgamated uses of methodologies in data science. The first section showed that the statistical components of data science are a necessary instrumentality for achieving its objectives of presenting information in a manner that is conducive to decision-making. The second section of the chapter showed that the selection of analytical pathways for business data is directly related to the goals of the organization or entity that is commissioning a practitioner. The third section showed that ML is one of the new pathways that are used to develop new products, services, and applications for data science. The fourth section laid out a business case for using data-driven science in all organizations, particularly those that seek to obtain a competitive advantage. The chapter closes with an overview of some of the conceptual aspects that together make up the empirical, theoretical, and computational underpinnings of data science. This chapter has shown that far from being a closed impractical endeavor, data science can have practical uses for all organizations. The next chapter emphasizes this by examining the changing faces of data science.

CHAPTER

3

THE CHANGING FACE OF DATA SCIENCE

CONTENTS

Data science is by definition very responsive to the changes that take place in the wider business environment (Davis et al., 2014). That means that the discipline has to overcome some of the reputational obstacles that it finds in its way as well as certain negative myths about the phenomenon which broadly fall under technophobia (Holmes, 2005). This chapter seeks to explore the ways in which data science is adjusting to the realities of the business world. The first section in this chapter will explore the ways in which technology was introduced and embedded in business. The second section will consider the implications of the data deluge. The third section focuses on database management techniques and approaches. The fourth section will consider the use of distributed and parallel systems in data analysis. The final section in this chapter will highlight the role of predictive modeling, business intelligence (BI), and analytics in the commercial and non-commercial corporate world today. This third chapter explores the themes of relevance that were introduced in the previous chapter, but at a less technical and more practical level.

3.1. INTRODUCTION OF INFORMATION TECHNOLOGY

The classic definitions of information technology or IT refer to it as the application of telecommunication equipment such as computers in order to store, transmit, retrieve, and manipulate data (Dutse, 2013). Typically, information technology will take place within the context of an organization (Hair, 2010). The modernization of management practices has meant that information technology is applicable in both profit and not-for-profit organizations (Howells and Wood, 1993). It is one of the important competencies for a manager regardless of whether they are operating in the public or private sector (McFarlane, 2010). Given the equipment that is associated with information technology, it is not surprising that the term has become almost synonymous with computer networks and hardware (Hamari et al., 2015). Nevertheless, it is important to consider the fact that information technology also includes other information distribution technologies such as telephones and televisions (Little, 2002). There are very many industries that are associated with information technology as their primary product (Jansen et al., 2008). Examples include computer hardware, software, electronics, semiconductors, internet, telecommunications equipment, engineering, healthcare, e-commerce, and computer services (Min et al., 2008). However, virtually all modern businesses use information technology and have therefore departmentalized the function (Min et al., 2009); either as a stand-alone department or one that has incorporated within production units (Miller, 2014).

The fact that information technology has exploded in the modern era can make it tempting to assume that this phenomenon is a new one (Engelberger, 1982). In reality, human beings have been engaging in some very basic computational technology as far back as 3000 BC (Hamari et al., 2015). However, the modern term for information technology came into being in 1958; courtesy of an article by Leavitt and Whisler in the Harvard Business Review (Little, 2002). They identified three broad categories of information technology (Helmreich, 2000). The first comprises the techniques that are used for processing information (Kim and Jeon, 2013). The section refers to the application of statistical or mathematical methods to the decision-making process (Little, 2002). The third is the simulation of higher-order thinking in non-animate objects such as computer programs and robots (McFarlane, 2010). Based on the storage and processing technologies employed, it is possible to distinguish four distinct phases of information technology development. For example, there is the pre-mechanical (3000 BC–1450 AD), mechanical (1450–1840), electromechanical (1840–1940) and electronic (1940–present). This is a discipline or area of scientific inquiry that shows every indication of continuing to grow with time (Hamari et al., 2015). Figure 3.1 highlights some of the key events in the development of information technology in modern history.

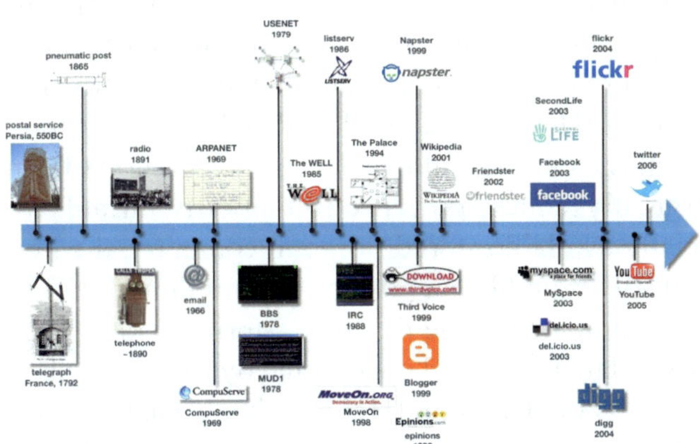

Figure 3.1. Brief modern history of information technology.

Source: Mohd Zharif.

The tally stick is a very early example of computational equipment (Boase, 2008). The Antikythera mechanism, dating from about the beginning of the first century BC, is generally considered to be the earliest known mechanical analog computer, and the earliest known geared mechanism (Hair, 2010). Comparable geared devices did not emerge in Europe until the 16th century (Hilbert and Lopez, 2011). In actuality, it was not until 1645 that the first mechanical calculator capable of performing the four basic arithmetical operations was developed (Kim and Jeon, 2013). Electronic computers, using either relays or valves, began to appear in the early 1940s (Bansal, 2013). The electromechanical Zuse Z3, completed in 1941, was the world's first programmable computer (Howells and Wood, 1993). Judging by modern standards, Zuse Z3 was one of the first machines that could be considered a complete computing machine (Howells and Wood, 1993). Colossus, developed during the Second World War to decrypt German messages was the first electronic digital computer (Ifinedo, 2016). Although it was programmable, it was not general-purpose, being designed to perform only a single task (Lyytinen et al., 2016). It also lacked the ability to store its program in memory (Ifinedo, 2016). Hence, programming at the time was carried out using plugs and switches to alter the internal wiring (Hilbert and Lopez, 2011). The first recognizably modern electronic digital stored-program computer was the Manchester small-scale experimental machine (SSEM), running its first program on 21 June 1948 (Kirchweger et al., 2015).

Even in the earliest days of developing information technology, there was concern about matching technical prowess with convenience for the use. Energy usage would be one of the ways in which the comfort of the user was addressed. More recently, this has turned into concern about the impact on the environment (Holmes, 2005). The development of transistors in the late 1940s at Bell Laboratories allowed a new generation of computers to be designed with greatly reduced power consumption (Kim and Jeon, 2013). The first commercially available stored-program computer, the Ferranti Mark I, contained 4050 valves and had a power consumption of 25 kilowatts (Kees et al., 2015). By comparison, the first transistorized computer, developed at the University of Manchester and operational by November 1953, consumed only 150 watts in its final version (Helmreich, 2000). Consumers were also demanding better services in terms of the processing and storage power of the computer (Holmes, 2005). In the early days of the computer, it was not uncommon to make use of punched tapes in order to represent data and that started the basis of working on the storage and processing power of computer devices (Min et al., 2009).

The punched tape that was used in the earliest electronic computers (e.g., Colossus) was a long strip of paper on which the data necessary for the functioning of the computer was represented as a series of holes (Hamari et al., 2015). Since World War II, there was an interest in using computers for data storage (Hilbert and Lopez, 2011). The early forms of this facility involved a form of delay line memory which was configured in such a way as to remove clutter from received radar signals (Ifinedo, 2016). The first practical manifestation of this development was the mercury delay line. The Williams Tube was the first random-access digital storage device (Ifinedo, 2016). This was based on a standard cathode ray tube (Jansen et al., 2008). The problem with the delay line memory and Williams Tube was that it was highly unstable. Regular updating is necessary in order to keep the information fairly intact (Ifinedo, 2016). Moreover, all the data was lost once power was removed (Mosher, 2013). By modern standards, that would be far an acceptable storage mechanism. The magnetic drum was the pioneer of non-volatile computer storage devices in 1932 (Howells and Wood, 1993). Hence, it was used in the Ferranti Mark 1 (Holmes, 2005). This was the world's first commercially available general-purpose electronic computer (Mieczakowski et al., 2011). IBM introduced the first hard disk drive in 1956, as a component of their 305 RAMAC computer system (Jansen et al., 2008). Most digital data today is still stored magnetically on hard disks, or optically on media such as CD-ROMs (Menke et al., 2007). Until 2002, most information was stored on analog devices, but that year digital storage capacity exceeded analog for the first time (Malathy and Kantha, 2013). As of 2007, almost 94% of the data stored worldwide was held digitally (Menke et al., 2007). This was broken down in terms of the 52% being stored on hard disks, 28% being stored on optical devices and 11% being stored on digital magnetic tape (Mieczakowski et al., 2011). It has been estimated that the worldwide capacity to store information on electronic devices grew from less than 3 Exabytes in 1986 to 295 Exabytes in 2007; hence doubling roughly every 3 years (McFarlane, 2010). This demonstrates the possibilities of modern digital devices and their contribution to information technology as a whole (Jansen et al., 2008).

Modern organizations are particularly interested in the development of databases that allow them to store, retrieve, and manipulate information long after it was collected (Hamari et al., 2015). During the 1960s, researchers started to explore the possibilities of database management systems (DBMS) (Kees et al., 2015). The reason for this was mainly due to the difficulties that had been encountered when storing and retrieving large amounts of sensitive information (Rachuri et al., 2010). The Information Management System or IMS was one of the earliest efforts by IBM to address their problem (Wallace, 2004). IMS has remained durable for more than 4 decades, albeit with some modifications which are designed to deal with the problems that were encountered along the way (Tarafdar et al., 2014). One of the key features of IMS is that it tends to store data hierarchically. Ted Codd proposed an alternative relational storage model in the 1970s (Sinclaire and Vogus, 2011). The proposal was underpinned by set theory and predicate logic (van Deursen et al., 2014). Therefore, this relational database would rely on concepts such as rows, tables, and columns (Wallace, 2004). The first commercially available relational database management system (RDBMS) was available from Oracle in 1980 (Hamari et al., 2015). All DBMS consist of a number of components that together allow the data they store to be accessed simultaneously by many users while maintaining its integrity (Ifinedo, 2016). A characteristic of all databases is that the structure of the data they contain is defined and stored separately from the data itself, in a database schema (Ruben and Lievrouw, n.d.). The extensible markup language (XML) has become a popular format for data representation in recent years (Hilbert and Lopez, 2011). Although XML data can be stored in normal file systems, it is commonly held in relational databases to take advantage of their robust implementation verified by years of both theoretical and practical effort (Jibril and Abdullah, 2013). As an evolution of the standard generalized markup language (SGML), XML's text-based structure offers the advantage of being both machine and human-readable (Jibril and Abdullah, 2013).

It is not enough to store data because a retrieval process is required in order to access that data for decision-making purposes (Abu-Saifan, 2012; Gibson and Brown, 2009; Sakuramoto, 2005). The relational database model introduced a programming-language independent structured query language (SQL). This is a language that is based on relational algebra (Bachman, 2013; Helmreich, 2000; McFarlane, 2010). The terms "data" and "information" are not synonymous, although they are sometimes used interchangeably by those that do not know any better (Davis et al., 2014; Min et al., 2008; Sobh and Perry, 2006). Anything stored is data (McFarlane, 2010), but it only becomes information when it is organized and presented meaningfully (Sakuramoto, 2005). Indeed, data is best used for decision-making when it has turned into information. Those that make decisions based on data because they have not bothered to engage in any kind of analysis are doing a disservice to their organization (Jansen et al., 2008). There is a fairly detailed and informed process that turns raw data into information that can well be used for purposes of making decisions (Ifinedo, 2016). Although this is a tedious and sometimes expensive process, it is also necessary in order to get the best out of the raw data (Ifinedo, 2016). The investment that is put into processing the data will more than likely be repaid by the quality of decisions that are made following the use of that refined data (Hilbert and Lopez, 2011). By refining, we do not mean hiding the key elements of the data simply because they do not make for pleasant reading (Menke et al., 2007). Rather, we mean being able to tease out the main and important themes that can turn that data into something on which business decisions can be based (Noughabi and Arghami, 2011).

Most of the world's digital data is unstructured, and stored in a variety of different physical formats even within a single organization (Davis et al., 2014). Data warehouses began to be developed in the 1980s to integrate these disparate stores (McFarlane, 2010). They typically contain data extracted from various sources, including external sources such as the Internet, organized in such a way as to facilitate decision support systems or DSS (Carr, 2010). The next issue for consideration was how and when data was going to be transmitted to the units where it would be used as information (Helmreich, 2000). Data transmission has three aspects: transmission, propagation, and reception (Min et al., 2009). It can be broadly categorized as broadcasting, in which information is transmitted unidirectional downstream, or telecommunications, with bidirectional upstream and downstream channels (Min et al., 2008). XML has been increasingly employed as a means of data interchange since the early 2000s, particularly for machine-oriented interactions such as those involved in web-oriented protocols such as SOAP, describing "data-in-transit rather than data-at-rest" (Ifinedo, 2016). One of the challenges of such usage is converting data from relational databases into XML document object model (DOM) structures (Miller, 2014). The data then has to be manipulated in order to address the research problems that were identified in the research brief (Min et al., 2009). This is what is known as data manipulation. Hilbert and Lopez identify the exponential pace of technological change (McFarlane, 2010). First, the machines' application-specific capacity to compute information per capita roughly doubled every 14 months between 1986 and 2007 (Mieczakowski et al., 2011). Secondly, the per capita capacity of the world's general-purpose computers doubled every 18 months during the same two decades (McFarlane, 2010). Third, the global telecommunication capacity per capita doubled every 34 months (Noughabi and Arghami, 2011). Fourth, the world's storage capacity per capita required roughly 40 months to double every 3 years (Menke et al., 2007). Fifth, the per capita broadcast information has doubled every 12.3 years (Mosher, 2013).

The age of databases is not without its critics (Awang et al., 2013; Hair, 2010; Mosher, 2013). Some are concerned about how the massive amounts of data being stored are going to be used and whether they may possibly fall into the wrong hands (Helmreich, 2000; Mieczakowski et al., 2011). Already consumers are experiencing the effects of push sales which do not have any consideration for their privacy but are instead obsessed with selling products and services to them at any cost (Wallace, 2004). Whereas a lot of data is collected and stored on a daily basis, not all of it becomes usable information (Hilbert and Lopez, 2011). The rest is stored in what are called data tombs (Hilbert and Lopez, 2011). These are those types of archives which are rarely visited by service users because they do not contain the kind of data that can easily be turned into useful information (Jansen et al., 2008). This is a source of worry for some consumers who believe that at some point, greedy corporations may reactivate this data and then use it in ways that are harmful to the consumer (Kees et al., 2015). Even more concerning, is the notion that the state can be an effective custodian of records given the poor history that the state generally has in terms of handling private information or even responding to the needs of private consumers (Ellison, 2004). During the 1980s, data mining became popular because it did not rely on the information stored in the data tombs (Helmreich, 2000). Instead, it focused on current and recurrent data from which it selected particular patterns that were of interest (Hilbert and Lopez, 2011). In that way, the data was being collected or analyzed for specific purposes (Lyytinen et al., 2016). However, data mining did not completely alleviate the concerns of consumers who suspected that they were constantly being monitored for purposes that were not expressly clear to them (Kim and Jeon, 2013). It was something of covert surveillance from which the private sector or any other vested interest could extract data at a time and in a manner of their choosing (Kirchweger et al., 2015). The risks could be incalculable and that has raised many consumer protection concerns (McFarlane, 2010).

The academic perspective also has some concerns about the collection of data without necessary ethical protections and informed consent (Gibson and Brown, 2009). The Association for Computing Machinery defines information technology as undergraduate degree programs that prepare students to meet the computer technology needs of business, government, healthcare, schools, and other kinds of organizations (Hilbert and Lopez, 2011). When performing these roles, the information technology specialists assume responsibility for selecting hardware and software products appropriate for an organization, integrating those products with organizational needs and infrastructure, and installing, customizing, and maintaining those applications for the organization's computer users (Holmes, 2005). However, such work is also undertaken with due diligence and reference to ethical standards which ensure that the subjects of data collection are protected (Hamari et al., 2015). Moreover, within academia; there are standard quality standards that must be maintained if any inferences are to be made from, the data that is collected (Menke et al., 2007). Unfortunately, the data deluge has meant that not enough attention is paid to the collection process (Stone et al., 2015). As a consequence, there is a lot of data out there without going through the proper procedures for collecting it (Gilks, 2016). That is why researchers are most interested in primary research because, at the very least, they are able to go back and check the data from the source (Rachuri et al., 2010).

The field of information ethics was established by mathematician Norbert Wiener in the 1940s (Malathy and Kantha, 2013). Some of the ethical issues associated with the use of information technology include breaches of copyright, China being a prime example of intellectual property laws that are too weak (Mosher, 2013). At the same time, too much restriction on copyright might mean that content that ought to be widely available is restricted (Min et al., 2009). Artists and innovators are particularly concerned that the fruits of their labor are being systematically stolen using the online medium (McFarlane, 2010). Indeed, there have been cases where copyrighted material is redistributed free of charge after paying an individual license fee (Noughabi and Arghami, 2011). Therefore, the public is already familiar with the content and may not see the need for paying for it again (Rachuri et al., 2010). There is also concern about surreptitious monitoring by people in authority such as employers and the state (Evans, 2009). Given the poor accountability that is already associated with modern politics, there is legitimate concern that private citizens will no longer be able to challenge their governments since the power relationships have been redefined by the data deluge (Howells and Wood, 1993). The collection of data is now enshrined in the law (Ifinedo, 2016). There are close circuit television cameras that are used to record virtually every aspect of our daily lives (McFarlane, 2010).

The problem of unsolicited communication from marketers has the power to derail the reputation of an organization and make it that much harder to gain voluntary informed consent to research (Ellison, 2004). However, privacy law campaigners argue that a lot of information gathering is not even after consent (Gibson and Brown, 2009). The information is collected as people use the internet or alternative request for services from companies that do not follow a strict data protection policy (Kees et al., 2015). Sometimes the information is not even shared with other third parties but is obtained through illegal means such as hacking (McFarlane, 2010). Recently, some major companies have been taken to court over breaches of client confidentiality and privacy after their databases were compromised by third parties (Min et al., 2009). At other times, information is corrupted after being collected which calls into question its veracity and relevance to the decision-making process (Min et al., 2009). Online databases are particularly vulnerable and that is why some customers make specific requests not to be included in such databases regardless of all the reassurances that are provided by the organization (Hair, 2010). Because breaches of privacy laws are actionable in courts, some organizations have erred on the side of caution and not engaged in any significant data collection regardless of the incentive to do so (Helmreich, 2000). Others have put stringent measures in place to ensure that they obtain informed consent before collecting data and that there are protections that prevent third parties from the access or abusing that data (McFarlane, 2010).

The other concern about the introduction of information technology has been the de-personalization of business (Bachman, 2013). Essentially, some businesses have lost the human touch because they are of the view that a lot can be gained from investing in computers that are capable of mimicking the performance of real human beings (Hair, 2010). This has led to an insidious process of de-skilling which robs the industry of some of its best workers who are then replaced by machines which are not always as effective in their roles (Lewis, 1996). A classic example of this taking place is how some public sector organizations are sending out very impersonal and hurtful letters to those who are already dead but who have not been accurately recorded on the system (Little, 2002). There was a notorious case in the UK where a woman's family received a notification from the welfare services team that they were stopping her benefits because she was dead. Although human error predates computer systems, it is hard to conceive that a worker reading messages would send out such an email or letter. Machines are not infallible and the age of information technology has sometimes given the erroneous impression that they are. When companies are making decisions, they can reference the input from their information technology departments (Jansen et al., 2008). However, it is imperative not to lose the human touch which creates a sense of judgment that cannot really be replicated by simple algorithms (Howells and Wood, 1993). The world of innovation has not yet come up with a satisfactory solution to the age-old question as to how human beings can be replaced in functional organizations (Lyytinen et al., 2016).

3.2. THE DATA DELUGE

We have already alluded to the data deluge and the kind of pressure that it can put on both the community at large and the corporate world (Bansal, 2013). The term data deluge has become something of a hackneyed refrain for everything to do with the difficulties of controlling how information is gathered, processed, and used (Gibson and Brown, 2009). Modern technology has been incorporated into virtually every aspect of our lives and it can be difficult for people to detach from its clutches even when they feel that it is eventually harming them in one way or another (Hamari et al., 2015). Those that have specialized in data science can collect information about virtually every aspect of our lives in order to persuade us to purchase their products or engage in other forms of desirable consumer behavior (Kirchweger et al., 2015). Sometimes, the very same organizations that are responsible for the explosion of the data deluge are the ones that complain about it (Little, 2002). Academia is increasingly interested in the nature and implications of the data deluge (Sinclaire and Vogus, 2011). At the same time, academia itself is subject to the data deluge and a perpetuator of the data deluge through its millions of research projects (Menke et al., 2007). In the defense of academia, the focus on ethical issues in research has meant that it has a much better record than the corporate world in terms of protecting the privacy of the participants in formal research activities (Mieczakowski et al., 2011). Figure 3.2 highlights the exponential demands of the data deluge.

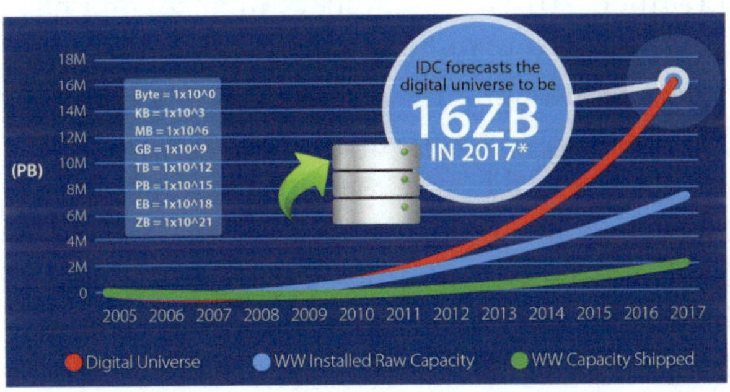

Figure 3.2. The data deluge.

Source: ASEAN Data and Storage.

The era of big data can have opportunities and challenges for the academic as well as the practitioner (Dutse, 2013). More important, it can also affect the decision-making process for all levels of all organizations (Hilbert and Lopez, 2011). In the early days, researchers would be able to analyze data that they had collected on a single computer and then share the results with the research team before publishing the findings without reference to the individual participants (Kees et al., 2015). The limited streams of information were controlled by the research aims, objectives, and hypotheses which were a condition of gaining ethical clearance to undertake research (Hamari et al., 2015). This controlled information flow could then be used to make decision in a specific context and with specific limits (Kim and Jeon, 2013). The researchers were at pains to explain the limitations of their research so that there was limited scope for exaggerating the importance or role of the data that they had acquired. Indeed, the peer review process ensured that any mistakes and omissions in the research process were critiqued (Malathy and Kantha, 2013). Those research projects that did not make the cut would not get ethical clearance (McFarlane, 2010). Certainly, the publishers would not release papers that were known to have ethical problems (Lyytinen et al., 2016). Those that were already in the public domain were withdrawn once the mistakes were uncovered (McFarlane, 2010).

The needs of industry have changed significantly and it is no longer sufficient to merely report what happened (Evans, 2009). The researcher must be able to explain why it happened and hopefully make some predictions about the future (Gilks, 2016). That means that the research has to be a lot more intensive and try to go behind the presenting phenomena (Helmreich, 2000). This might mean collecting even more data and going into more detail about that data (Miller, 2014). Researchers in academia have tried to meet the needs of industry whilst maintaining their ethical standards (Lyytinen et al., 2016). This can be a tough task and there is a mixed record on success (Schute, 2013). Technology is very much part and parcel of these changes because it allows for the rapid processing of large amounts of data (Gibson and Brown, 2009). This technology might not make a distinction between following carefully designed scientific experimentation and unstructured data that is mined from online sites (Evans, 2009). One only has to think about the challenges that faced researchers in the era before packages such as SPSS, Epidata, and Atlas. The concurrence of scientific advancement and technological advancement has been a boom for academia, but it has also led to challenges such as maintaining the highest ethical standards when dealing with very large volumes of data from diverse populations (Jibril and

Abdullah, 2013). Others argue that the use of technology has created a fallacy that mistakes association with causation (Lyytinen et al., 2016).In an effort to get quick and marketable results, researchers are increasing exaggerating the significance of their findings (Kobie, 2015). Often, researchers have been accused of misleading people with contradictory information which is sometimes published simultaneously (Miller, 2014). A classic case in point is the science about the role of carbohydrates in a healthy diet. Researchers have veered from vilification of carbs to their vindication to a cautious recommendation of their inclusion in the diet. Someone reading such contradictory research may not be able to make a decision either way.

The data deluge has become so important that it has led to the development of careers that would have been unthinkable a few years back (Kobie, 2015). Analysts are focusing on many niches which change the ways in which we view life and its effects on the wider environment (Holmes, 2005). Managers are being educated about the possible uses and mishaps that can arise when working with big data (Menke et al., 2007). Meanwhile, some of the negative assumptions about the data deluge are being challenged and firms are finding new ways of making use of big data (McFarlane, 2010).

3.3. DATABASE MANAGEMENT TECHNIQUES

Given the increasing importance of databases, it is not surprising that enterprises are investing in developing techniques for managing them better (Berker et al., 2006). The new database management techniques will help to organize big data into information as and when required by various actors within industry (Hilbert and Lopez, 2011). The modern era means that all businesses, regardless of their size or scope, must have some element of data management (Hilbert and Lopez, 2011). Given the sensitivity of this aspect of the business and the significant training requirements for staff, there are companies that have invested in software such as a DBMS (Jansen et al., 2008). This can be an off-the-shelf product or one that is made to meet the specifications that are given out by the company (Holmes, 2005). The important thing is to ensure that a high quality of data access is guaranteed for both internal and external actors (Howells and Wood, 1993). All businesses, from mom-and-pop storefronts to global corporations, need to manage data (Hilbert and Lopez, 2011). For decades, DBMS have served as important method of data management (Min et al., 2009). DBMS software automatically stores and organizes data and protects it with sophisticated security (Little, 2002). Today, several different types of databases are available on computing devices ranging from smartphones to large-scale mainframes (Mieczakowski et al., 2011). Figure 3.3 illustrates some of the constituent elements of DBMS.

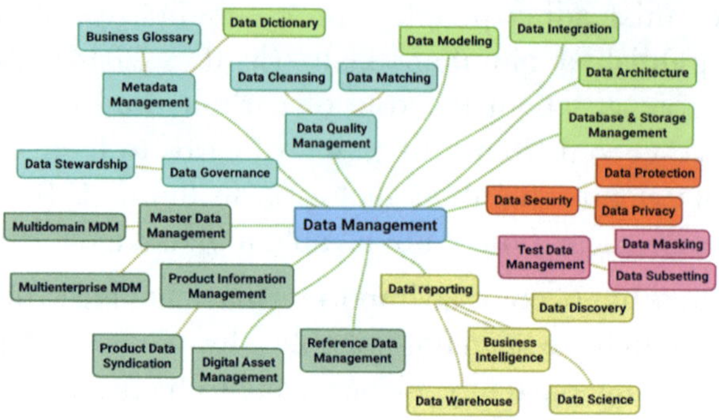

Figure 3.3. Data management in an organization.

Source: Lilien Dahl.

Databases are known as repositories of large amounts of data (Awang et al., 2013). They are different from a single file or a small group of files (Hair, 2010). The advantage of using databases is the fact that they are quite fast and can process information at a pace that is suitable for the business needs (Hilbert and Lopez, 2011). A case in point is the databases that hold customer accounts in banks and service providers (Howells and Wood, 1993). The government also maintains large databases about its citizens and other citizens (Malathy and Kantha, 2013). The other advantage of databases is that they can be configured in such a way as to ensure that the data which they contain is secure and is not compromised by third parties (Awang et al., 2013). This is particularly important for businesses that may use information as a competitive advantage, in which case they would not like third parties to gain access to their business intelligence (BI) (Hamari et al., 2015). The client-server model is one of the successful additions to database management techniques in as far as it allows for multiple accesses whilst maintaining control over the data itself (Gilks, 2016). There will be a program that is known as the service (Holmes, 2005). This program is responsible for controlling all aspects of the database including storage protocols, organization of data and access issues (Howells and Wood, 1993). For example, a person or entity cannot gain access to the database content without inputting the correct usernames and passwords (Kim and Jeon, 2013). Moreover, the database maintains a record of access for audit purposes and only gives access to the information that is allowed and has been requested by the user (Min et al., 2009). The client is a separate program that makes requests to the server (Bachman, 2013). Although you can have a system consisting of only one client and one server, it is more common to have multiple clients and one server, or multiple clients and servers (Noughabi and Arghami, 2011).

Database server software protects data with sophisticated security (Dutse, 2013; Hilbert and Lopez, 2011). You can assign different passwords to users or user groups, and allow different levels of security for different types of data (Helmreich, 2000). In addition, most databases let you create rules that automatically check your data as you add and update it, helping to minimize errors (Kim and Jeon, 2013). For example, you can create a rule for the state in an address, that it must be a correct U.S. state abbreviation. The protocols that are assigned to the database play an important role in ensuring that it is capable of handling the data requests from the users. Indeed, databases are able to handle such complex issues such as the travel movements of all users of an international airport. Quite often, the focus is on ensuring that the database is secure and that third parties cannot penetrate it in order to change the protocols that were originally assigned to it (Howells and Wood, 1993). One of the tricks that are used by hostile competitors is to corrupt a database using malware which sends out instructions that are in contravention of the original purpose of the database (Little, 2002). That is why the most effective database management techniques also create provisions for ensuring a good level of security for both the people that access the database and those that own it (Mieczakowski et al., 2011). There are a number of database types that require slightly different management techniques depending on the circumstances:

- **Type 1 – Relational Databases:** A common type of database is the relational kind, which can temporarily join data stored in separate tables into complex and useful combinations (Hamari et al., 2015). For example, you can connect salesperson, customer, and order data to create a list of salespeople and their biggest customers (Hilbert and Lopez, 2011). Although some databases store data as numbers, such as dollar amounts or dates, and text, such as name and address (Lewis, 1996); others can store video, documents, and other complex information (Mieczakowski et al., 2011). Earlier network and hierarchical databases were similar to the relational kind but less flexible in terms of combining data (McFarlane, 2010).

- **Type 2 – Database Design and Organization:** Database design is a task typically undertaken by a software developer, app designer or database specialist (Hilbert and Lopez, 2011). Whatever their background, this professional compiles a list of requirements from the users and creates the database organized into functional parts (Gilks, 2016). Examples may include things such as separate data tables for customers (Howells and Wood, 1993), inventory management (Helmreich, 2000), and order data protocols (Holmes, 2005). The specialist may also add some required security (Jansen et al., 2008), data-checking rules (Howells and Wood, 1993), and programming code which typically helps to build the apps that work with the database (Holmes, 2005).
- **Type 3 – Structured Query Language (SQL):** Many relational databases use the SQL. This is an English-like programming language used to create, maintain, and retrieve data from the database (Gilks, 2016). In the client-server model, a client app creates a message in the form of a SQL command and sends it to the server (Helmreich, 2000). The server receives the message (Hilbert and Lopez, 2011), checks its security passwords (Hamari et al., 2015), executes it, and passes the results back to the client (Kees et al., 2015). A simple SQL command to list customer information from the database might look like this: "SELECT name, address, city, state, zip-code FROM customer-table ORDER BY name."

In addition, the organization may decide to take advantage of the many pre-configured data management tools (Hair, 2010). These could be sold as part of an overall software package or they could be added gradually as the need arises (Hilbert and Lopez, 2011). Currently, it is common practice to make use of apps that can execute certain commands following a given query by the user (Gilks, 2016). Generally speaking, database management techniques are meant to be widespread within the workforce so that employees can use the facility of data systems to support their own work (Jansen et al., 2008). Part of the training is to be able to access the database accurately and obtain the information that is required for their work (Engelberger, 1982). The other part will be about having ethical considerations when accessing the database. For example, the fact that there are data protection laws will be of importance to those workers that deal with sensitive information concerning members of the public whose privacy must be maintained (Bachman, 2013). A case in point is where a media story will run a story and people in the tax office may access their records without authorization. This can lead to serious legal implications including imprisonment and significant fines for the organization that has failed to protect the public (Gilks, 2016). The teams that use the database must also be trained on how to recognize unusual usage patterns which may be an indicator of an even bigger problem that must be addressed as a matter of urgency (Kim and Jeon, 2013). This is important when engaging in fraud prevention strategies or any other preventative measures that are meant to ensure that all the stakeholders in the organization are full protected against digitized attacks (Ifinedo, 2016).

3.4. DISTRIBUTED AND PARALLEL SYSTEMS

The operation of distribution and parallel systems is an important consideration of data science (Hair, 2010). Indeed, they are so important that each has led to the development of a distinct sub-field within the wider discipline (Hair, 2010). Hence, we speak of distributed computing which focuses on distributed systems whose components are to be found in different computers that happen to be networked (Kirchweger et al., 2015). The computers are able to communicate with each other through the passing of messages (Sin, 2016). Although they are separate computers, they are united by a common link and purpose (Sakuramoto, 2005). Distributed systems have been associated with three characteristics that distinguish them from their parallel counterparts (Menke et al., 2007). First, all the components are concurrent or simultaneous. Secondly, there is no global clock (Schute, 2013). Thirdly, there is an independent failure of components (Min et al., 2009). Examples of distributed systems vary from SOA-based systems to massively multiplayer online games to peer-to-peer applications (Min et al., 2008). A computer program that runs within a distributed system is called a distributed program (Dutse, 2013). Moreover, distributed programming is the process of writing such programs (Holmes, 2005). There are many different types of implementations for the message passing mechanism (Hamari et al., 2015; Miller, 2014; van Deursen et al., 2014). These include pure HTTP (Jansen et al., 2008), RPC-like connectors (McFarlane, 2010), and message queues (Min et al., 2009). Figure 3.4 is a comparative analysis of parallel and distributed systems.

Parallel v.s. Distributed Systems

	Parallel Systems	Distributed Systems
Memory	**Tightly coupled shared** memory UMA, NUMA	**Distributed** memory Message passing, RPC, and/or used of distributed shared memory
Control	**Global clock** control SIMD, MIMD	**No global clock** control Synchronization algorithms needed
Processor interconnection	Order of **Tbps** Bus, mesh, tree, mesh of tree, and hypercube (-related) network	Order of **Gbps** Ethernet(bus), token ring and SCI (ring), myrinet(switching network)
Main focus	**Performance** Scientific computing	Performance(**cost and scalability**) **Reliability/availability** **Information/resource sharing**

Figure 3.4. Distributed and parallel computer systems.

Source: Computer Systems Textbook.

Distributed computing also refers to the use of distributed systems to solve computational problems (Carlson, 1995). In distributed computing, a problem is divided into many tasks, each of which is solved by one or more computers (Howells and Wood, 1993). These computers communicate with each other via message passing frameworks (Jansen et al., 2008). The word parallel in terms such as "parallel system," "distributed programming," and "distributed algorithm" originally referred to computer networks where individual computers were physically distributed within some geographical area (Awang et al., 2013; Hair, 2010; Hilbert and Lopez, 2011). Currently, these terms are used in their wider sense because they refer to all the autonomous processes that can run on a single physical computer whereby there is interaction with other computers through the mechanism of message passing (Hilbert and Lopez, 2011).

There is no complete and exclusive definition of a distributed system in existing literature today (Helmreich, 2000). However, the researcher has identified a number of properties that are associated with distributed systems and which they utilize in order to loosely define the term (Jansen et al., 2008). First, there will be many autonomous computational entities which are actually computers or nodes (Lewis, 1996). Each of these will have their own localized memory (Menke et al., 2007). Message passing is then used to communicate with the various computing entities (Miller, 2014). There may be a common goal that unites the distributed systems (Bansal, 2013). For example, the entire network could be set up in order to resolve a significant computational problem (Lyytinen et al., 2016). In doing so, the user is able to conceive and perceive the collection of all the autonomous processors as a single unit (Little, 2002). Another approach is where each computer has its own user with certain user-defined tasks and needs (Malathy and Kantha, 2013). Therefore, in this case, the role of the distributed system is to ensure proper coordination of the shared resources as well as an avenue for communication with other users of the system (Min et al., 2008). The distributed system has to be configured in such a way as to tolerate failures by the individual computers so that the other computing entities can continue working (Gibson and Brown, 2009). There is a structure to the system which includes network latency, network typologies, and the number of computing entities (Holmes, 2005). However, this structure is not known in advance (Howells and Wood, 1993). That means that the system may comprise different network links and computers which can change during the execution of a given distributed program (Mieczakowski et al., 2011).

Distributed systems are groups of networked computers, which have the same goal for their work (Ifinedo, 2016). The terms "concurrent computing," "parallel computing," and "distributed computing" have a lot of overlap, and no clear distinction exists between them (Hamari et al., 2015). The same system may be characterized both as "parallel" and "distributed" (Jibril and Abdullah, 2013). However, the processors in a typical distributed system run concurrently in parallel (Jibril and Abdullah, 2013). Parallel computing may be seen as a particular tightly coupled form of distributed computing and distributed computing may be seen as a loosely coupled form of parallel computing (McFarlane, 2010). Some of the key differences that have been identified in existing literature help to distinguish parallel from distributed computer systems (Helmreich, 2000). For example, in parallel computing, all processors may have access to a shared memory to exchange information between processors (Holmes, 2005). In distributed computing, each processor has its own private memory or distributed memory (Kobie, 2015). Information is exchanged by passing messages between the processors (Lewis, 1996). The situation is further complicated by the traditional uses of the terms parallel and distributed algorithm that do not quite match the above definitions of parallel and distributed systems (Little, 2002). Nevertheless, as a rule of thumb; high-performance parallel computation in a shared-memory multiprocessor uses parallel algorithms while the coordination of a large-scale distributed system uses distributed algorithms (Kirchweger et al., 2015).

The history of the differentiation of parallel and distributed computer systems really started in the 1960s when there was customary use of concurrent processes which could then communicate using message passing

(Gilks, 2016). The distributed systems were then made available to the wider public on a commercial basis with products such as Ethernet which came into being in the 1970s (Howells and Wood, 1993). Towards the end of the 1960s, the ARPANET was introduced and this, in fact, became the father of the internet (Ifinedo, 2016). By the beginning of the 1970s, the world had been introduced to the ARPANET email service and this became one of the most successful applications (Kobie, 2015). Some researchers argue that this remains the first example of large-scale distributed application (Kirchweger et al., 2015). By the 1980s, there were other networks such as FidoNet and Usenet (Holmes, 2005). The study of distributed computing became its own branch of computer science in the late 1970s and early 1980s (Menke et al., 2007). The first conference in the field, symposium on principles of distributed computing (PODC), dates back to 1982 (Mieczakowski et al., 2011). The counterpart International Symposium on Distributed Computing (DISC) was first held in Ottawa in 1985 as the International Workshop on Distributed Algorithms on Graphs (Min et al., 2008).

From a business decision-making perspective, it is important to have clarity about the kind of architecture that is used for both the parallel and distributed computer systems (Chesbrough, 2005). There are many options that are made available thanks to the advances in technology and the responsiveness to consumer demands (Hamari et al., 2015). For example, at the lower level of operations; the multiple central processing units (CPU) might be interconnected into a network (Holmes, 2005). This would be the case regardless of whether the network has been printed onto a circuit board or whether it comprises of loosely coupled devices (Jibril and Abdullah, 2013). When it comes to the higher level, it may be appropriate to interconnect the various processing that are running on the CPU using a communication system (Malathy and Kantha, 2013). The more bespoke this communication is, the better it is for purposes of operational functionality (Spiekermann et al., 2010). Distributed programming typically falls into one of several basic architectures. These include the client-server (Howells and Wood, 1993), three-tier (Helmreich, 2000), n-tier (Kim and Jeon, 2013), and peer-to-peer (Jansen et al., 2008). They can also be categorized as loose coupling or tight coupling (Kees et al., 2015).

1. **Client-Server:** Architectures where smart clients contact the server for data then format and display it to the users (Hamari et al., 2015). Input at the client is committed back to the server when it represents a permanent change (Ifinedo, 2016).

2. **Three-Tier:** Architectures that move the client intelligence to a middle-tier so that stateless clients can be used (Ifinedo, 2016). This simplifies application deployment (Ifinedo, 2016). Most web applications are three-tier (Mieczakowski et al., 2011).
3. **n-Tier:** Architectures that refer typically to web applications which further forward their requests to other enterprise services (Lyytinen et al., 2016). This type of application is the one most responsible for the success of application servers (Lewis, 1996).
4. **Peer-to-Peer:** Architectures where there are no special machines that provide a service or manage the network resources (Helmreich, 2000). Instead, all responsibilities are uniformly divided among all machines which are known as peers (Howells and Wood, 1993). Peers can serve both as clients and as servers (McFarlane, 2010).

Another basic aspect of distributed computing architecture is the method of communicating and coordinating work among concurrent processes (Howells and Wood, 1993). Through various message passing protocols, processes may communicate directly with one another (Little, 2002). This is typically in a master or slave relationship (Mosher, 2013). Alternatively, a "database-centric" architecture can enable distributed computing to be done without any form of direct inter-process communication (Miller, 2014). This is achieved practically by utilizing a shared database (Min et al., 2008). There are many reasons why an organization may adopt distributed computing and systems (Chiu et al., 2016). For example, the application itself may be configured in such a way that it requires a communication network which links into various computers (Holmes, 2005). This is the case where data is produced in one physical location has to be transferred to another location in order to be worked on by the employees (Hamari et al., 2015). At other times, the use of a single computer may not be possible regardless of the preferences of the organization (Lewis, 1996). These are cases where a distribution system is deemed to be more practical (Miller, 2014). A case in point is where the company decides that it can save costs and achieve its objectives by linking together several low-end computers as opposed to purchasing a single high-end computer (Miller, 2014). The distributed system itself has advantages which make it more appropriate for some organization (Hair, 2010). For example, research has shown that a distributed system is more reliable than its non-distributed counterparts (Hilbert and Lopez, 2011). This is because there is no single point of failure and therefore activities can continue even when one computing entity has broken down (Howells and Wood, 1993). Another advantage is that the distributed system makes it easier to engage in expansion projects when compared to a monolithic uniprocessor system (Hilbert and Lopez, 2011). Therefore, organizations that constantly have to upgrade their systems might prefer the distributed option (Helmreich, 2000).

There are many examples of industry applications of distributed systems including: telecommunication networks (Carlson, 1995); telephone networks and cellular networks (Jansen et al., 2008); computer networks such as the Internet (Ellison, 2004); wireless sensor networks (Little, 2002); routing algorithms (Ifinedo, 2016); network applications (Ifinedo, 2016); World Wide Web and peer-to-peer networks (Noughabi and Arghami, 2011); massively multiplayer online games and virtual reality communities (Jansen et al., 2008); distributed databases and distributed DBMS; network file systems (Jansen et al., 2008); distributed information processing systems such as banking systems and airline reservation systems (Ruben and Lievrouw, n.d.); real-time process control (Mieczakowski et al., 2011); aircraft control systems (Min et al., 2009); industrial control systems (Ruben and Lievrouw, n.d.); parallel computation (Berker et al., 2006); scientific computing including cluster computing and grid computing (Jibril and Abdullah, 2013); various volunteer computing projects (Lewis, 1996); and distributed rendering in computer graphics (Min et al., 2009). The wide applications of these systems represent their utility in business (Helmreich, 2000).

3.5. BUSINESS ANALYTICS (BA), INTELLIGENCE, AND PREDICTIVE MODELING

Some of the critical process for applying data science includes BI, business analytics (BA), and predictive modeling (Carr, 2010). There is a difference between business and predictive analytics, despite the variety of references that can become a bit confusing for those that are not steeped in the technology tradition (Howells and Wood, 1993). Although some technologies might appear to be replicating duties, they in reality have different functions that together contribute to the business development that is a necessity for industry to prosper (Howells and Wood, 1993). Companies cannot always distinguish between BA and predictive modeling (Gilks, 2016). It is only after education and experience that they can make the distinctions (Gilks, 2016). These distinctions, in turn, help organizations to make the best use out of BA and predictive modeling (Helmreich, 2000). The primary purpose of BA is to enable an organization to make better decisions (Engelberger, 1982). This occurs after leveraging many tools and methodologies (Hamari et al., 2015). Therefore, BA necessarily involves data mining, data analytics, and big data functions which are all geared towards more appropriate decision-making (Kees et al., 2015). In fact, the new developments in the sector mean that the tools that are used for BA can generation their own reporting and visuals without the constant supervision of personnel (McFarlane, 2010). Figure 3.5 highlights the continuum of analytical capabilities for a typical business, with BA and predictive modeling being part of that progression over time.

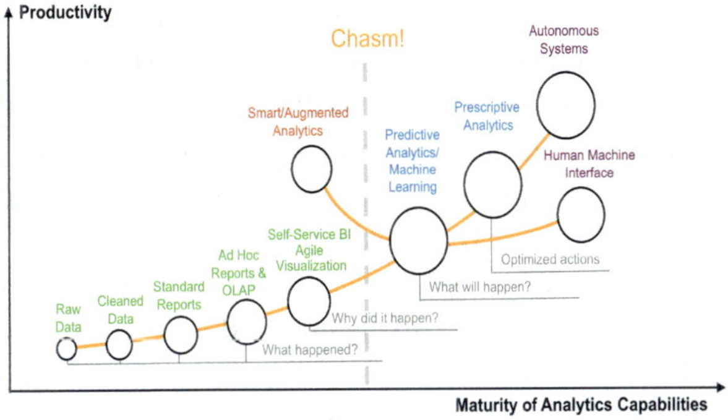

Figure 3.5. Progression in analytical capabilities.

Sources: Timo Elliot.

As its name suggests, predictive analytics is all about attempting to understand what the future will be like based on what is happening today and what happened in the past (Awang et al., 2013). This is an important business asset because it means that the decisions which organizations make are proactive and preventative rather than being reactive and curative (Evans, 2009). All things being constant, proactive, and preventative action costs much less to the organization than situations where it has to react to problems and then try to deal with them after they have occurred (Hamari et al., 2015). That is why many organizations today are heavily investing in predictive analytics as a means of improving their efficiency (Holmes, 2005). Information technology has meant that it is possible to predict trends based on big data and high-level analysis (Malathy and Kantha, 2013). Predictive Analytics is a hot issue in today's business and information technology world (Hamari et al., 2015). Predictive analytics goes beyond these backward-facing views and uses the data you already hold in your business to look forward and tell you what is going to happen in the future

(Holmes, 2005). Moreover, predictive modeling can provide businesses with information about the next best thing that could happen in the future (Holmes, 2005; Mosher, 2013; van Deursen et al., 2014). The best predictive analytics tools will automate this process so that business decision making becomes fact-based and truly data-driven rather than based on subjective judgments and hunches (Dutse, 2013; Kees et al., 2015; Ulloth, 1992).

There is very practical and useful information that can be gleaned from predictive modeling (Awang et al., 2013). For example, you may be able to identify the products that are doing well on the market and those that are underperforming (Chesbrough, 2005). This, in turn, will inform your decisions about the priority investment areas and even the personnel decisions that you may have to take in the wake of all the changes that are taking place (Hamari et al., 2015). The predictive model can chart the product life cycle so that you can change your marketing strategies accordingly in order to avoid being caught off guard by trends (Little, 2002). The outcomes of a predictive modeling exercise may include the launch of a specific advertising campaign that is informed by the information that you have received about the relationships between and among the variables that affect your product on the market (Mieczakowski et al., 2011). It can give you valuable insights into consumer behavior and how that consumer behavior is likely to change over time (Min et al., 2008).

In order to ensure the best possible accuracy, the predictive analysis uses various models to analyze data (Awang et al., 2013). The most common is the predictive model algorithm that is focused on the individual customer behavior (Hair, 2010). Using sample data with known attributes, the model is trained and is able to analyze the new data and determine its behavior (Hilbert and Lopez, 2011). This information can be used to predict how the customer might behave next (Mosher, 2013). One of the key concerns when relying on predictive modeling is the extent to which the predictions are accurate (Mieczakowski et al., 2011). Specifically, the information that is relied on to make the prediction may be faulty in one way or another (Mosher, 2013). Moreover, the person doing the predictive analysis may not be experienced enough to pick out the relevant themes that can make a difference to their performance (Menke et al., 2007). All these are considerations that must be given their due weight when selecting an analytical approach (Sakuramoto, 2005). It is also imperative that the business is not over-reliant on predictive modeling to the extent that it forgets the practice-based knowledge that arises with experience in the industry (Hilbert and Lopez, 2011). The entire premise of handling big data must be dynamic in order to account for the dynamic nature of the information that is being dealt with (Hamari et al., 2015). There are no straightforward solutions to an industry that is quite complex and might not even follow linearity, particular with regards to consumer behavior (Holmes, 2005). The company must therefore have a contingency plan that is designed to deal with those events that are unexpected or are enacted in ways that divert away from the norm (Malathy and Kantha, 2013). The predictive analysis is supposed to be a buffer against uncertainty but it does not completely eliminate the possibility of uncertainty (Miller, 2014).

CHAPTER 3: SUMMARY

This chapter sought to highlight the changing face of data science in order to identify the ways in which this particular aspect of decision-making is relevant to modern organizations. The first section showed that information technology has coincided with an increased awareness of the need to combine objective and subjective conceptualizations of business success. Specifically, there is a demand for data-driven decision making which is considered to be more sustainable and less prone to instability. In thc second section of this chapter, we saw that the exponential accumulation of information about people has led to the data deluge which can be confusing for those organizations that are looking for the main issues that need urgent attention in their business environment. The third section explained some of the database management techniques that are currently popular in the corporate world and the rationale for their popularity. In the fourth section, we distinguished between distributed and parallel computer system; an important aspect of information technology architecture. The last section showed that predictive modeling and BA can form a powerful mix for BI as long as corporations and other interested entities are aware of their limitations. The next chapter will consider the statistical applications of data science.

CHAPTER 4

STATISTICAL APPLICATIONS OF DATA SCIENCE

CONTENTS

The previous chapter showed how data science is adapting to the realities of a business world that makes extraordinary demands on it. This chapter goes into a rather technical application in the form of the use of statistical data in a range of contexts and for a range of purposes. The first section of this chapter considers the use of data science in the public sector. The second section discusses how data science can become a competitive advantage in industry. The third sector explores some data engineering practices. In the fourth section, we look at applied data science in action. The chapter closes with a review of the predictive and explanatory theories of data science. The primary purpose of this chapter is to open up the possibilities of using data science beyond its technical concerns.

4.1. PUBLIC SECTOR USES OF DATA SCIENCE

The public sector has been accused of being very poor at taking up data science (Bansal, 2013). However, such accusations do not account for the fact that the public sector has many rules and procedures which are meant to ensure that it is accountable to the citizenry of its location (Gibson and Brown, 2009). Therefore, public sector data science is not only of interest to academics but also industry leaders (Kees et al., 2015). Sometimes the role of public sector data science is explicitly located within the vision and mission statement of each team (Hair, 2010). However, at other times it is merely implied by virtue of the goals of that department (Miller, 2014). In some ways, public sector data science is an instrumentality in terms of helping to deliver better and more appropriate public services (Helmreich, 2000). For the public sector, data science can be a very critical resource when trying to identify service users, understand their needs, and engage with them in a meaningful, but respectful manner (Davis et al., 2014). Unlike the private sector that wishes to sell products for a profit, the public sector is looking to ensure that the state and community at large fulfills its duty of care to the citizenry (Holmes, 2005). The priorities are different from the private sector, but public sector data science is no less professional than other alternative forms of analytics (Helmreich, 2000). Indeed, the performance management requirements of the public sector today mean that there is an increased attention on the ability to use data responsibly but effectively (Little, 2002). Nowhere is this more evident than in the implementation of preventative healthcare where the service units seek to identify people at risk and help them to lead healthier lifestyles in order to prevent them from having to go to hospital when it is too late (Howells and Wood, 1993).

The public sector works with a variety of clients and will, therefore, have a variety of information technology needs (Cappellin and Wink, 2009). Data science tries to rationalize these interactions so that they are optimized and provide value for money with regards to taxpayer funds (Jansen et al., 2008). Given the high accountability requirements for the public sector, it is also imperative to ensure that the data science which is being used to make business decisions is as robust as possible (Kirchweger et al., 2015). For example, data science can help public sector organization to find sources of funding and people in need (Lyytinen et al., 2016). It can also form the basis of mutually beneficial strategic partnerships which can reduce the burden on the public sector (Malathy and Kantha, 2013).

Using this science will also be helpful when trying to detect abuse of the system such as benefit claimants who are also working or who are not being truthful about their needs assessment (Helmreich, 2000).

The bureaucratic standards that are encountered during the commissioning of information technology might mean that the public sector is not as flexible as it ought to be when dealing with such large scale projects (Jibril and Abdullah, 2013). As a consequence, there is a tendency to purchase old equipment that is not up to the kinds of standards that service users might expect (Jansen et al., 2008).

Besides, the long history of poor performance in the public sector and the fact that there is no pressure to make sales might lead to complacency so that poorly functioning systems are eventually accepted as a way of life in the public sector (Ellison, 2004).

Some employees are so entrenched in the idiosyncrasies of the public sector that they are unwilling to engage in any new system, no matter efficient and attractive it might be (Jansen et al., 2008). Short risking a public backlash by terminating the employment contracts of these workers, the department heads have no alternative but to moderate their update of information technology in order to accommodate entrenched workers (Mosher, 2013). Figure 4.1 highlights some of the challenges and opportunities that data science represents for the public sector.

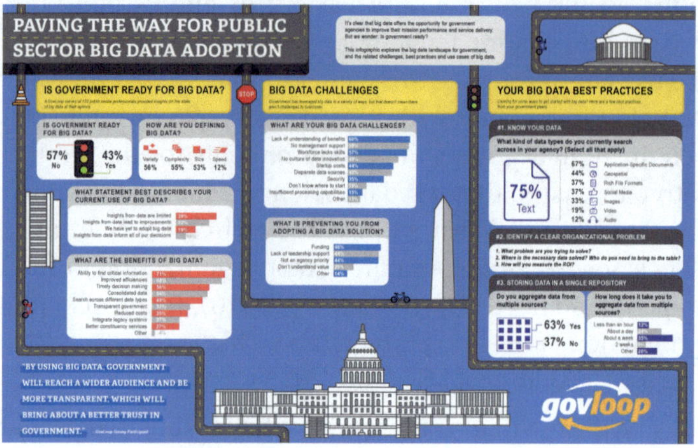

Figure 4.1. Data science in the public sector.

Source: Visually.

Because the government has an obligation to cater for all its citizens who are in need, the inclusion, or targeting criterion for service delivery is not as stringent as that of the private sector (Gilks, 2016). Whereas the private sector focuses on ability and willingness to pay, the public sector focuses on need within a priority framework. Regardless of the different focus points, both sectors benefit from the insights that data science can provide during the decision-making process (Howells and Wood, 1993). Moreover, public sector data science helps to set performance measures that can help to rationalize practices across different department as well as ensuring that the public is fully engaged in the service delivery (Ellison, 2004). Government is sometimes obliged to engage with clients who are hard to reach or otherwise disengaged from the service provision (Helmreich, 2000). The business intelligence (BI) that arises out of public sector data science can be a useful tool for understanding what is going well and what needs to be improved in order to achieve the goals of a given team (Min et al., 2009). The public sector also has a strong collaborative framework that includes several departments that may each contribute in a unique way towards the provision of a specific service (Hamari et al., 2015). The value of public sector data science is in being able to provide a way of communicating as well as the material that needs to be communicated so that the decisions that are taken by each team are rational and within the overall plan of the organization (Helmreich, 2000).

The complexity of public sector data science means that quite often the support framework is outsourced to a private company in order to allow the government to continue engaging in its public service ethos without worrying too much about commercial concerns (Chiu et al., 2016). For example, the public sector data science may focus on promoting a safe immunization program for new mothers in deprived communities. The contractor will get some baseline data and also some statistics about the behavior of the targeted consumers (Jibril and Abdullah, 2013). This will then inform government policy and will be translated into the specific actions of the units or teams that are supposed to ensure that the immunization program takes place (Holmes, 2005). The government structures and institutions can be vast in their scope and complexity (Engelberger, 1982). A case in point is the USA federal government structures which sit atop state authorities and are responsible for coordinating a myriad of activities that involve members of the public (Holmes, 2005). Perhaps one of the biggest public sector data science project is that which relates to taxes and welfare payments (Jibril and Abdullah, 2013). Virtually every citizen will have record in one or more of the databases (Helmreich, 2000). The task is in ensuring that these databases effectively and efficiently communicate with each other so that they do not harm the service delivery process (McFarlane, 2010). A case in point is where clients get irritated when they are being asked for the same information over and over again because the departments cannot communicate with each other or when serious criminal escape justice because the various police agencies are not sharing intelligence (Mieczakowski et al., 2011).

Some of the possibilities of public sector data science are best understood with specific programs that have successfully supported public sector work (Bansal, 2013). A case in point is the Civis program which is used in the USA (Davis et al., 2014). The primary objective of Civis is to find populations at risk, or those that require outreach (Hilbert and Lopez, 2011). Civis empowers public sector organizations to change the way they interact with citizens, increasing their power to make substantive change across communities. Civis specifically focuses on identifying key populations of interest, and determining which interventions are most likely to impact those populations. For example, we are helping local government and community-based organizations in New York City locate residents that are likely eligible for benefits like SNAP or the earned income tax credit (EITC), but are not accessing those programs. There are many lessons that can be gleaned from those public sector organizations that have made use of data science (Engelberger, 1982). First of all, it is imperative to have a lot of collaboration between and among departments in order to improve the outcomes for the citizenry (Hair, 2010). Data analytics traditionally focuses on understanding what has happened in the past by finding trends and patterns in historical data to inform future decisions (Mieczakowski et al., 2011). Data science focuses on the application of artificial intelligence and machine learning (ML) to identify future outcomes (Chiu et al., 2016). Public sector organizations must not assume that they are exempt from the need to keep up with the times and identify ways of working that might improve their efficiency and effectiveness (Little, 2002). Together, data analytics and data science can reveal valuable insights across the vast and expanding lake of public sector data (Berker et al., 2006). The responsibilities of the department do not just end at making decisions, but they also doing due diligence in order to ensure that the data collection process does not infringe on the rights and expectations of the public at large (Holmes, 2005). Once the output from public sector data science has been curated, it is important to ensure that it is stored securely and only used according to the implicit or explicit agreements made with the public with regards to the issue of responsible data usage (Little, 2002).

Public sector innovators are coming up with new and interesting ways of interacting with customers (Awang et al., 2013). A case in point is the use of analytics in order to assess the length and quality of voice calls as well as social media messages (Gilks, 2016). In this way, it is possible to construct a framework that monitors customer sentiments in real-time and then compare them with other time frames in order to understand how consumer attitudes and behavior are changing (Kees et al., 2015). There have been instances where customer complaints have been halved as a consequence of appropriately applying public sector data science (Gibson and Brown, 2009). Knowing what the customer wants can reduce the stress and pressure that public sector workers face when they are trying to decipher complex problems that the community brings to them (Little, 2002). ML has been successfully used in order to accurately predict the likelihood that certain customers will become brand ambassadors for services that are provided by the public sector (Bachman, 2013). This, in turn, can improve the outreach programs which sometimes meet resistance from the public due to negative stereotypes about the quality of public services (Little, 2002). It is also possible to counter certain false narratives about public services which are peddled by politicians for purposes of persuading potential voters to their side (Noughabi and Arghami, 2011). In order to maximize the benefits that arise out of public sector data science, it is important to take certain critical steps and corrective measures towards these goals (Engelberger, 1982). First of all, the ethos of the public service must shift away from merely acquiring technology and towards ensuring positive outcomes for service users (Helmreich, 2000). Hence, it is no longer about the technical capabilities of the systems such as database management techniques, APIs, and fancy algorithms (Ifinedo, 2016). Rather, the debate turns to the experiences of people who for one reason or another are utilizing public services (Jibril and Abdullah, 2013). The ideal is where public sector data science is better aligned to service delivery, instead of getting overly concerned about the technical accomplishments that are associated with particular systems (Miller, 2014). These are the kinds of public sector data science projects that are focused on the consumer who in this case is a member of the public (Helmreich, 2000).

4.2. DATA AS A COMPETITIVE ADVANTAGE

The competitive advantages of data science have been explored in existing literature (Chiu et al., 2016). Organizations are constantly trying to win new customers without losing their hold on their existing customers (Helmreich, 2000). Those organizations that know how to leverage the competitive advantages of data science have been able to distinguish themselves from the rest of the pack (Jibril and Abdullah, 2013). According to the McKinsey Global Institute (MGI), data-driven organizations are now 23 times more likely to acquire customers, 6 times as likely to retain customers, and 19 times as likely to be profitable as a result (Chiu et al., 2016). These are statistics that should be of concern to any organization that is still living a past where data was not that important and gut instinct is the primary decision-making determinant (Hamari et al., 2015).

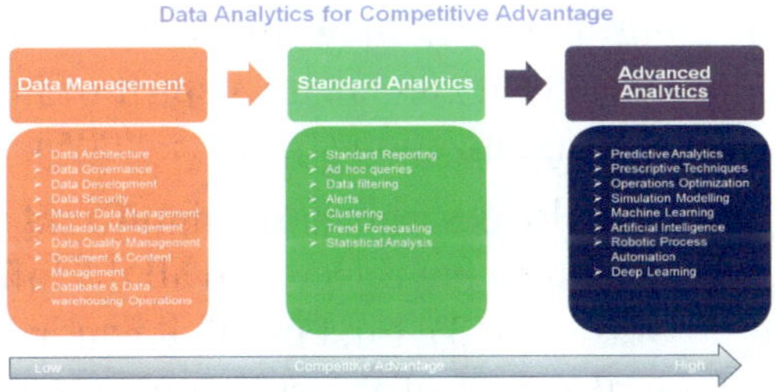

Figure 4.2. Data analytics as a competitive advantage.

Source: Arun Kottolli.

Even those organizations that used to get away with instinctive decisions are now realizing that they cannot sustain their success with such an approach without considering the possibilities of data (Jansen et al., 2008). Figure 4.2 highlights some of the competitive advantages of data science with specific reference to data analytics.

Another advantage of relying on objective data is that it allows for the successful transition of the organization from one phase to another without significant disruptions (Bansal, 2013). The systems and BI within the organization can be maintained even when the lead participants or employees are no longer with the organization (Gibson and Brown, 2009). That means that those organizations that are able to successfully leverage the competitive advantages of data science are able to maintain a good level of consistency over the long run (Hair, 2010). They are not subject to fluctuations in performance based on the probability of having good or bad luck (Kim and Jeon, 2013). Companies that wish to achieve longevity and sustainability in their operations should leverage the competitive advantages of data science as much as possible (Engelberger, 1982). Using data allows everyone to understand what they are doing and why they are doing it (Helmreich, 2000). By understanding the implications of their decisions, managers are more prudent and will ensure that they are acting in the interests of the organization over the long run (Helmreich, 2000). Data allows the players in an organization to understand what is happening in the internal and external environment (Ifinedo, 2016). Because this data is provided on a real-time basis, they are not caught off guard by trends that start occurring unexpectedly (Ifinedo, 2016). Through leveraging the competitive advantages of data science, it is possible to ensure accountability and transparency across the organization (Ulloth, 1992). Everyone knows that the others are doing and how they affect the immediate operations (Howells and Wood, 1993). That can create a spirit of healthy competition that ultimately benefits the organization (Holmes, 2005).

The response from employees can have significant benefits from the organization if they feel that they have access to all the information that they need to do their work (Dutse, 2013). First of all, it can increase engagement because the workers are also part of the solution to the problems that are identified (Hamari et al., 2015). When sales are down, it is not just the concern of the organization or its senior executives (Howells and Wood, 1993). Even the lowliest support employee will take an interest because the information chain has added them to the loop (Min et al., 2008). Even more importantly, data science can provide employees with possibilities of helping to intervene in order to deal with a corporate problem (Lewis, 1996). For example, if the organization is on the brink of bankruptcy; it may so happen that all departments start to cut back on their expenditure in an effort to ensure that this eventually does not occur (Howells and Wood, 1993). This is very different from an organization where everything is done in secret by the senior executives (Jibril and Abdullah, 2013). That means that the average employee will either not buy into the strategic plan or will actively buy out from that strategic plan (Malathy and Kantha, 2013). Others may seek to sabotage the company because they feel that they are not part of the positive aspects that are taking place in that organization (Wallace, 2004).

The fact that information is provided readily means that there is an impetus to take responsibility and act (Engelberger, 1982). The organization as a whole can become very proactive because decisions are made on the latest information which is carefully curated and organized so that it has relevance to the priorities of the organization (Dutse, 2013). Employees are part of the solution and will ideate certain ideas that can change the organization towards a better orientation based on the feedback that is coming from the environment (Hair, 2010). Over time, this will mean that the brand will construct a reputation for resilience and responsiveness (Little, 2002). Consumers are always enthusiastic about brands that seem to go that extra mile in order to deal with the emerging needs and expectations of their clients (Kees et al., 2015). Data-driven organizations have the additional advantage of obtaining feedback from their customers about how they are doing and some of the measures that they could take in order to make their services even more appealing to their clients (Holmes, 2005). This is very different from those reactive organizations that only deal with customer issues when there is a serious complaint that threatens to destroy their reputation within the industry (Howells and Wood, 1993).

Leveraging competitive advantages of data science means that the organization can make fast and confident decisions that can be defended in any arena and in any situation (Ifinedo, 2016). This is different from those organizations that take decisions that are not based on defensible and therefore the organization is constantly second-guessing itself (Ifinedo, 2016). When organizations are making decisions based on gut instinct, it can take some time before the decision-maker is comfortable making the call (Holmes, 2005). This is particularly true if they are self-conscious enough to understand that they could be subject to biases which render their decision inappropriate (Ifinedo, 2016). However, if the decision is made based on objective data; things become relatively quicker and straightforward (Jansen et al., 2008). These efficiencies and the decisiveness of the organization contributes to its relationship with customers who are more likely to stick with the brand because it has values, but at the same time is able to adjust its performance depending on the feedback that is being sent out by the customers (Ifinedo, 2016).

Data are now woven into every sector and function in the global economy (Bansal, 2013; Helmreich, 2000). Just like other essential factors of production such as hard assets and human capital, much of modern economic activity simply could not take place without them (Chesbrough, 2005; Holmes, 2005; Kim and Jeon, 2013). The use of Big Data will become the basis of competition and growth for individual firms (Chesbrough, 2005; Engelberger, 1982; Howells and Wood, 1993). These are large pools of data that can be brought together and analyzed to discern patterns and make better decisions (Davis et al., 2014; Helmreich, 2000; Mosher, 2013). This offers a number of advantages including enhancing productivity and creating significant value for the world economy (Bansal, 2013; Chiu et al., 2016; Hamari et al., 2015). This is achieved by reducing waste and increasing the quality of products and services (Bachman, 2013; Dutse, 2013).

Until now, the torrent of data flooding our world has been a phenomenon that probably only excited a few data geeks (Bachman, 2013; Hamari et al., 2015; Helmreich, 2000). But we are now at an inflection point. According to research from the MGI and McKinsey and Company's Business Technology Office, the sheer volume of data generated, stored, and mined for insights has become economically relevant to businesses, government, and consumers (Bansal, 2013; Boase, 2008; Engelberger, 1982; Hilbert and Lopez, 2011). The history of previous trends in IT investment and innovation and its impact on competitiveness and productivity strongly suggest that Big Data can have a similar power (Bansal, 2013; Jansen et al., 2008; Carr, 2010). This is in effect the ability to transform our lives (Helmreich, 2000). The same preconditions that allowed previous waves of IT-enabled innovation to power productivity are in place for Big Data (Bachman, 2013; Ellison, 2004; Holmes, 2005). These include technology innovations followed by the adoption of complementary management innovations (Davis et al., 2014; Hamari et al., 2015; McFarlane, 2010). Consequently, we expect suppliers of Big Data technology and advanced analytic capabilities to have at least as much ongoing impact on productivity as suppliers of other kinds of technology (Dutse, 2013; Hilbert and Lopez, 2011; Mieczakowski et al., 2011). All companies need to take Big Data and its potential to create value seriously if they want to compete (Bachman, 2013; Hamari et al., 2015; Jibril and Abdullah, 2013). For example, some retailers embracing big data see the potential to increase their operating margins by 60% (Hair, 2010; Sakuramoto, 2005; Zhang and Chen, 2015). The companies that will benefit most from the competitive advantages of data science are the ones that recognize the limitations of the old approach (Awang et al., 2013; Hair, 2010).

4.3. DATA ENGINEERING PRACTICES

Data engineering is a technical process that is responsible for ensuring that the benefits of data science are fully experienced within the organization (Bachman, 2013). We already know that data drives most of the business activities in our contemporary world (Gibson and Brown, 2009). Typically, the organization that is thinking of adopting a data-driven approach will have a number of business questions and problems that need to be resolved satisfactorily before that organization can take its place within the industry (Helmreich, 2000). One of the questions will seek to understand where the business growth is and what is it worth to acquire one more customer in a given segment (Hamari et al., 2015). The company will be looking at possible improvements and then subject them to a cost-benefit analysis in order to assess whether it is worth their while to engage in such activities (Abu-Saifan, 2012). Ideally, the organization will engage in the production of output that is most appealing to the consumers that they are targeting (Engelberger, 1982). The adoption of different data engineering business practices is meant to ensure that the answers to these questions are put into practice (McFarlane, 2010). Figure 4.3 highlights some of the modern data engineering business practices.

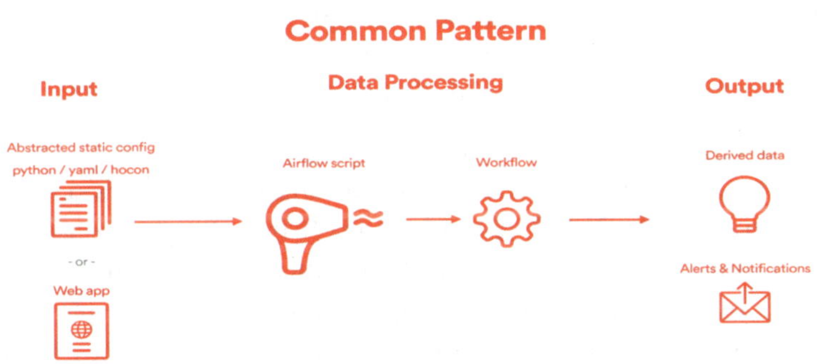

Figure 4.3. Data engineering business practices.

Source: Medium.

One of the ways in which data engineering business practices are enacted in an organization is through the development of different systems which end up becoming data production units (Awang et al., 2013). The data is the narrative of all the activities that take place within that system as well as the environment dynamics that affect that narrative (Gilks, 2016). Therefore, data engineering business practices are meant to ensure that there is a way of constructing and understanding a narrative of what happens in any given production or operational unit within an organization (Ifinedo, 2016). We may, for example, consider how the customer service system can start generating data for the organization (Hair, 2010). First, there will be the account information for each customer including their name, address, and biographical details (Helmreich, 2000). This is then supplemented by a database of their ordering and payment activities which includes shipments, cancellations, and orders (Jibril and Abdullah, 2013). Another system may consider their relationship with the organization in terms of any customer care complaints that they may have raised (Howells and Wood, 1993). There may yet be another database of psychometric properties which maps consumer behavior, bearing in mind the fact that the customer may have alternative concerns and possible connections with other organizations that provide the same or similar output when compared to the company that is undertaking the data analysis (Howells and Wood, 1993). When all this data is aggregated, the business will have a well-rounded picture of the customer so that they can provide appropriate services to them (Davis et al., 2014). It can also be of benefit to develop a sustainability framework to ensure that the customer will remain with the company over the long run (Hamari et al., 2015). The data sets that comprise this record are independent of each other, but data engineering business practices ensure that they are interlinked and those relationships are clearly mapped out in a way that can help to make decisions (Noughabi and Arghami, 2011).

Without appropriate protocols that underpin data engineering business practices, it becomes difficult to answer challenging questions about consumer behavior (Davis et al., 2014). Already, it is clear that those organizations that are able to understand and use the data will be at a significant advantage when compared to those that have a much more lackluster performance on this issue (Gilks, 2016). The challenges of managing data can affect any company at any level. We know that even the smallest companies have an enormous amount of data to contend with and this data can be stored in very large repositories (Lyytinen et al., 2016). The interlinkages can become so complex that they overwhelm the system and will, therefore, lead to a breakdown if there is no sufficient infrastructure to support them (Menke et al., 2007). One of the key roles of the data engineering business practices is to ensure that there is facilitation for analysis (van Deursen et al., 2014). This facilitation makes life easy for the data scientists, analysts, and executives that need to make decisions based on that data (Zhang and Chen, 2015). The output from the data engineering business practices must be reliable, fast, and secure so that it provides the optimum support to the decision-maker (Helmreich, 2000). Data engineering must source (Hamari et al., 2015), transform (Ifinedo, 2016), and analyze data from each system (Mosher, 2013). For example, data stored in a relational database is managed as tables, like an Excel spreadsheet (Kim and Jeon, 2013). Each table contains many rows, and all rows have the same columns (Mosher, 2013). A given piece of information, such as a customer order, may be stored across dozens of tables (Menke et al., 2007).

There are other approaches to data engineering business practices which are dependent on the operational requirements of each unit that is working on the issues (Hair, 2010). For example, data stored in a NoSQL database such as MongoDB is managed as documents, which are more like Word documents (Hilbert and Lopez, 2011). Each document is flexible and may contain a different set of attributes (Kees et al., 2015). When querying the relational database, a data engineer would use SQL (Howells and Wood, 1993). This is different from MongoDB which has a proprietary language that is very different from SQL (Ruben and Lievrouw, n.d.). Data engineering works with both types of systems, as well as many others, to make it easier for consumers of the data to use all the data together, without mastering all the intricacies of each technology (Miller, 2014; Ulloth, 1992; van Deursen et al., 2014; van Nederpelt and Daas, 2012). The take-home from such requirements activities is the fact that even the simplest questions may require significant data engineering business practice changes in order to find the right answers (Davis et al., 2014). In order to optimally work with each system, the decision-maker must have some understanding of the data and the technology that they will be dealing with (Hair, 2010). This understanding can be achieved through specific training, development, and coaching (Holmes, 2005). Others may also supplement their competencies through practice knowledge (McFarlane, 2010). Experience can be a great asset to the decision-maker and that is why it is imperative to allow members of staff to engage with the data engineering business practices on a regular basis (Rachuri et al., 2010). Often, setting up the system is the hardest task and things become considerably easier when all is said and done.

The trends and history of data engineering business practices is a long one and which reflects the changing priorities for the main actors within the sector (Dutse, 2013). As companies become more reliant on data, the importance of data engineering continues to grow (Holmes, 2005). Since 2012, Google searches for the phrase have tripled for example (Chiu et al., 2016). There is an increased awareness about the potential of data and the various ways that it can be used in order to strengthen the position of various businesses within a given industry (Hair, 2010). The sheer volume and complexity of Google searchers indicate that even consumers are looking for information as they make their purchasing their decisions (Jibril and Abdullah, 2013). If laypeople can take the effort to search for data before making decisions, what about organizations that have entire departments which are dedicated to research and development?

It is not just purchasing decisions that are being done after online searches (Ellison, 2004). We know that there is an exponential increase in job searches on an annual basis and that this trend intensifies during period of economic instability (Helmreich, 2000).

Companies in the modern era must take an interest in the search trends that relate to data engineering because it can be one of the ways in which best practice can be shared more widely (Helmreich, 2000). Besides, the fact that data is now a valuable resource means that companies are very protective of it (Gibson and Brown, 2009). Accessing the best data may require more expenditure and other resources (Kim and Jeon, 2013).

Companies are finding more ways to benefit from data (Berker et al., 2006; Jansen et al., 2008; Lyytinen et al., 2016).

They use data to understand the current state of the business (Abu-Saifan, 2012; Berker et al., 2006; Mosher, 2013). They are also using data to predict the future (Howells and Wood, 1993), model their customers (Ifinedo, 2016), prevent threats (Sin, 2016), and create new kinds of products (Noughabi and Arghami, 2011).

Even as the data itself and the process for collecting it becomes complicated, the data engineering business practices must focus on ensuring that the ultimate output is easy to understand and use (Engelberger, 1982). It is important to acknowledge the need to break down data into components that make it easier for the decision-making process

(Howells and Wood, 1993). Eventually, data science will cease to be the exclusive domain of technical people, but a concern that touches on every segment of the organization (Lewis, 1996). Indeed, those that are put in positions which require decision-making will make an effort to ensure that they rely on data since they will have known from experience the kinds of benefits that that approach can bring (Lewis, 1996).

4.4. APPLIED DATA SCIENCE

The interest in applied data science is based on the kinds of benefits that multiple corporations have received from using it as the basis of decision making (Chiu et al., 2016). There is some skepticism about applied data science in general, with some wondering whether it is a passing fad that will not transform the industry over the long run while others consider it to be a once-in-a-lifetime opportunity that can change the future prospects of a business if it is handled well in the present (Evans, 2009). Existing literature has espoused a range of perspectives concerning the utility of applied data science, with varying degrees of veracity (Jibril and Abdullah, 2013). However, there is no doubt that this remains one of the major forces in the industry today (Kees et al., 2015). The benefits of applied data science speak for themselves and they are the justifications for implementing the various programs that are meant to enhance data-based decision making (Berker et al., 2006). There are different levels of application from the most basic to the most advanced (Gibson and Brown, 2009). The business or decision-maker will determine at which level they wish to conduct their analysis and application (Ifinedo, 2016). Moreover, applied data science has a bright future that includes many as-yet-unknown elements that should help businesses significantly if they position themselves as data-driven organizations (Malathy and Kantha, 2013). Figure 4.4 highlights some aspects of applied data science in a range of domains.

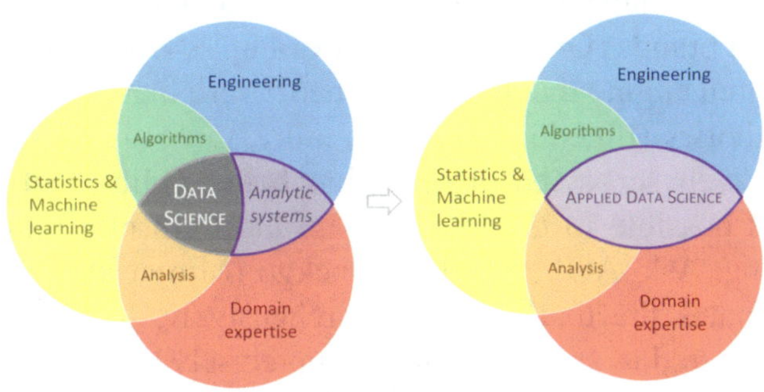

Figure 4.4. Application of data science in various domains.

Source: Utrecht University.

It is not just a question of having access to data science, but also being able to use it intelligently (Ellison, 2004). Part of this might be to carefully select those situations that require reliance on data science while utilizing some of the other capabilities that the organization has in different situations (Helmreich, 2000). For example, a database may be able to predict consumer behavior, but all that information is nearly useless if there is no customer relationship representative to build up the connections that acquire, service, and maintain a customer (Howells and Wood, 1993). The unintelligent use of data imagines that it is the only solution to every problem that the organization is facing (Kobie, 2015). Besides, the data that is collected is not only designed to be kept for a rainy day, but must be actively utilized before it begins to lose its relevance (Gilks, 2016). Intelligence use data analysis to know about which mix of variables is most likely to bring out the best in a business situation (Howells and Wood, 1993). Some companies also make the mistake of neglecting the issue of execution (Hair, 2010). It all very well knows what needs to be done, but you will only get results if you are able to actually do the things that are required (Ifinedo, 2016). It is not enough to profess knowledge about consumers if you are not taking active steps to address some of the issues that those consumers are raising about your output (Jibril and Abdullah, 2013). Remember that consumers are human beings with their own perspectives and behavior which may be so complex that you cannot reduce it to a few formulas that are presented in a data pack (Holmes, 2005). The organization that is going to dominate the future will be engaged in a constant search for knowledge in order to cement and expand its current situation (Kim and Jeon, 2013).

Over time, organizations have restricted themselves to a small menu through the new socialization and the successful advertising campaigns by some dominant brands (Gilks, 2016). For example, Google has been such a successful search engine that it has spawned a verb that seeks to encompass all search activities (Ifinedo, 2016). In reality, there are many alternatives which could even serve the organization better (Min et al., 2009). For example, it is possible to get localized results from other providers such as Yahoo, Bing, Ask, AOL, and Duckduckgo (Min et al., 2008). Because Google dominates the market so comprehensively, these smaller search engines have started to specialize and it is precisely that specialization that you may be looking for (McFarlane, 2010). If you can create a niche for yourself, it might be a better fit than if you are competing with many other players on a giant search engine (McFarlane, 2010). Remember that Google on average processes more than 20 petabytes worth of data on a single day

(Miller, 2014). That is a lot of processing, but it is also processing that is rather generic (McFarlane, 2010). You as the service provider may be looking for localized searches that are visited by a select group of people that may even have passed the first ring of inclusion by signing up to the search engine (Holmes, 2005). If you have, the product that they are looking for, your path towards making a full purchase will be much easier than if you are flooding the internet with content about products that millions of other providers can offer; sometimes at even better terms (McFarlane, 2010). The important thing to take away is that you have to open your horizons rather than restricting them to the most aggressive purveyor of a specific type of applied data science (Gilks, 2016). The savvy consumers are doing it and there is no reason why equally savvy businesses cannot go down the same route (Hilbert and Lopez, 2011).

It is also useful to be able to recognize data science when you see it (Helmreich, 2000). Some companies are so focused on the bottom line that they do not take the time to truly understand their environment and what it means (Howells and Wood, 1993). Initially, data is represented as a sea of details that may not make particular sense in your situation (Menke et al., 2007). The savvy entrepreneur will be able to curate that data and select the trends that are relevant to the field that they wish to explore (Kees et al., 2015). They will then turn that data into highly sophisticated decision-making frameworks that are based on evidence as well as detailed analysis (Jibril and Abdullah, 2013). Let us take the example of digital advertisements that may be released by competitors (Engelberger, 1982). It makes sense for a business that is seeking to penetrate a particular market to try and understand how the more experienced firms have been doing their business (Hair, 2010). The entrepreneur should know the differences between a targeted advertising campaign and one which re-targets consumers that have already interacted with the brand in some way (Kees et al., 2015). There is an entire spectrum of digital activities that go into making this advertising campaign work (Lyytinen et al., 2016).

All of them are driven by data science as well as the more niche field of ML (Jibril and Abdullah, 2013). The challenge for the entrepreneur, who wishes to make inroads, is to be able to map out how the entire campaign has been put together (Dutse, 2013). They will be looking at strengths which they can mimic and enhance in their own campaign (Ifinedo, 2016). They may also be looking for weaknesses that need to be mitigated in order to provide their own campaign with the best chance of success in the future (Kobie, 2015). Obviously, this is highly specialized work that will require the services of a dedicated team that can tease out all the aspects that need to be addressed (Noughabi and Arghami, 2011).

Beyond the complex algorithms that eventually display on the screen as an advertisement, there are real human efforts and relationships that must be included in the calculus about what is going to take place in a specific campaign (Hair, 2010). It is because of advanced data science that digital ads are able to attract a lot of attention and conversions by a click when compared to traditional advertising (Helmreich, 2000). These are digital adverts that have been carefully calibrated based on the data that is available and high-level analysis that can support the decision-makers in terms of the design and execution of the final display for the consumer (Ifinedo, 2016). The advert will also have gleaned a lot of information about consumer behavior which is then manifested by the ways in which the campaign is eventually allowed to progress (Spiekermann et al., 2010). Sometimes the targeting is so detailed that the advertising is personalized to individuals who are most likely to purchase the product (Gibson and Brown, 2009). This type of specialization is not possible without applied data science (Hilbert and Lopez, 2011). Another innovation that is being used for marketing purposes is the implementation of a recommender system for consumers (Little, 2002). When people make purchases, the predictive analytics will tell the seller about the other complementary or supplementary products that might be of interest to the person (Kirchweger et al., 2015). Therefore, they receive a short notification about the availability of these products on the premise that they are more likely than the average visitor to purchase the recommended products (Berker et al., 2006). Tracking consumer behavior can be controversial because it appears to be a very intrusive form of big data (Lyytinen et al., 2016). However, many organizations are exploring the possibilities of this type of applied data science (Sobh and Perry, 2006).

Image recognition has emerged as a potentially controversial but very useful form of applied data science (Gilks, 2016). We are beginning to train machines to recognize faces and that means that a lot of the surveillance that use to be undertaken by law enforcement officers can now be passed on to machines within an acceptable tolerance of error (Ifinedo, 2016). For example, close circuit television cameras are being updated with artificial intelligence that recognizes faces of known offenders who have outstanding warrants (Hilbert and Lopez, 2011). That means that they can easily be apprehended for walking down the street as opposed to the traditional searches that were once undertaken by law enforcement officers (Malathy and Kantha, 2013). Perhaps the problem with this particular type of technology is emblematic of the wider problems that are associated with applied data science (Lyytinen et al., 2016). When the face recognition cameras were put in place, there were trial runs that showed a very high error rate (Malathy and Kantha, 2013). This is an error rate that the public considers to be unacceptable, even for those people that have a criminal record (Carlson, 1995). Some worry that a government that is capable of monitoring every aspect of our lives is one that is too powerful to be held accountable (Kirchweger et al., 2015). Private companies are also using this technology in ways that could be harmful to private individuals (Gilks, 2016). For example, a user can upload their image with friends on Facebook and then start getting suggestions to tag known and assumed friends (Zhang and Chen, 2015). This automatic tag suggestion feature uses face recognition algorithm. However, some of the people that are tagged may no longer be in touch or willing to be friends with the person (van Deursen et al., 2014). Similarly, while using WhatsApp web, you scan a barcode in your web browser using your mobile phone (Davis et al., 2014). Google provides you the option to search for images by uploading those (Chiu et al., 2016). It uses image recognition and provides related search results. All these are useful developments, but one which also have serious ethical implications (Bansal, 2013; Min et al., 2009).

Other applications that have gained in popularity include speech recognition software which can be used to improve the social functionality of people with speech impairment or other related disability (Hilbert and Lopez, 2011). It can also be linked to other password-protected systems, therefore reducing the time that it takes to complete an authentication procedure (Mosher, 2013). Some of the best examples of speech recognition products is Google Voice, Siri, and Cortana which are very popular with young executives (Hamari et al., 2015). Those who are unable to or unwilling to type down text can still use technology through speech recognition. In that sense, this type of applied data science is helping to expand access (Malathy and Kantha, 2013). Of course, those that have used speech recognition fully understand the fact that it can never completely replicate the sophistication of a human brain. That is why some really strange translations can take place when relying on this type of technology (Lewis, 1996). Despite some misgivings, the applied data science relating to consumer interactions with technology have increased (Hilbert and Lopez, 2011). Nowhere is this more felt than in the gaming industry. EA Sports, Zynga, Sony, Nintendo, and Activision-Blizzard have led gaming experience to the next level using data science (Miller, 2014). Games are now designed using ML algorithms which improve and upgrade themselves as the player moves up to a higher level (Mieczakowski et al., 2011). Motion games allow for comparative analysis of competitor performance in order to elicit a competent response from the current player (Jansen et al., 2008). Similarly, price comparison websites are constantly looking for ways to incorporate applied data science into their own operations (Menke et al., 2007). At a basic level, these websites are being driven by lots and lots of data which is fetched using APIs and RSS Feeds (Jansen et al., 2008). PriceGrabber, PriceRunner, Junglee, Shopzilla, and DealTime are some examples of price comparison websites (Noughabi and Arghami, 2011).

4.5. PREDICTIVE AND EXPLANATORY THEORIES OF DATA SCIENCE

The theoretical framework that underpins explanatory and predictive data analysis is well studied, but not yet fully understood (Dutse, 2013). There has been such an effort to ensure proper execution of new technologies in real-time that sometimes the predictive and explanatory theories of data science have been neglected (Hair, 2010). This is based on the misguided view that data science is all about output and that to spend so much time trying to understand the underlying theory is a waste of time (Hilbert and Lopez, 2011). Some attention has been given to the technical aspects of data science including its sub-fields of predictive modeling, data mining, and ML (Helmreich, 2000). Others have researched the contemporary manifestation of data science and how they link back to its history (Mieczakowski et al., 2011). Because this is a relatively new aspect of business management, some of the theoretical foundations are borrowed from existing theories which are then amalgamated into a body of work that seeks to understand the role of data science in modern industry (Sakuramoto, 2005). Figure 4.5 highlights the fundamentals of the predictive and explanatory theories of data science as well as their linkages to the technical aspects and processes of this field.

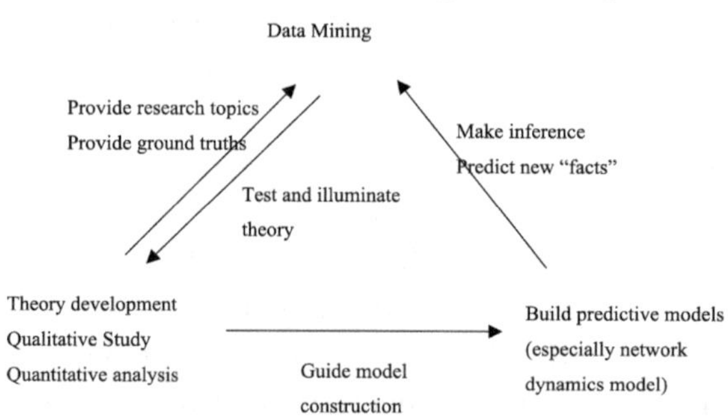

Figure 4.5. Predictive and explanatory theories of data science.

Source: Luo et al. (2016)

The business functionality of predictive models is to exploit knowledge about the present and past in order to predict the future (Ellison, 2004). The historical and transactional data may be available to many businesses, but it is only the data-driven ones that will take an interest in the patterns that can lay the foundation for predicting what might happen in the future (Helmreich, 2000). The ideal is where the business or decision-maker knows everything that there is to know about the future of their interests (Engelberger, 1982). However, that is not a perfect scenario that is ever achieved (Spiekermann et al., 2010). Instead, the business will have varying degrees of accuracy based on its approximation of the future (Lyytinen et al., 2016). The advantage for these businesses is the ability to capture and understand the relationships between the variables in order to develop a model of consumer and producer behavior (Howells and Wood, 1993). In business, predictive models exploit patterns found in historical and transactional data to identify risks and opportunities (Kirchweger et al., 2015). Models capture relationships among many factors to allow assessment of risk or potential associated with a particular set of conditions, guiding decision-making for candidate transactions (Micczakowski et al., 2011). The level of specificity in the information that is provided may render it limited to a particular firm in particular circumstances (Hamari et al., 2015). However, generalizability has not always been an important consideration in predictive and explanatory theories of data science since the aim is to ensure that the best information for a particular decision is available (Jansen et al., 2008). For the more general outlook, you have to go back to the raw data that has not yet been through the analytical processes (Miller, 2014).

The defining functional effect of these technical approaches is that predictive analytics provides a predictive score (probability) for each individual element in the model. This element could be a customer, employee, healthcare patient, product SKU, vehicle, component, machine, or other organizational units (Awang et al., 2013; Evans, 2009; Howells and Wood, 1993; Wallace, 2004). This information is provided in order to determine, inform, or influence organizational processes that pertain across large numbers of individuals, such as in marketing, credit risk assessment, fraud detection, manufacturing, healthcare, and government operations including law enforcement (Bachman, 2013; Hair, 2010; Ifinedo, 2016; McFarlane, 2010). The simplicity of the model means that it can be used in different industries as long as sufficient account is given for the changes in the variables which are bespoke to that particular industry (Hamari et al., 2015). We can look at the example of credit scoring which is not yet fully understood by the people that are subject to it. The bank or lender has no way of accurately knowing what the person is going to do when they get a loan. Therefore, they rely on the list of probabilities that calculate the possibility or likelihood of defaulting. Hence, someone that has a history of defaulting on loans with high commitments, low income, pending debts, a criminal record, and no citizenship has a very likelihood of defaulting than someone who has always paid their debts on time, has a steady job, has no criminal record, and is a permanent resident of the jurisdiction. It is not really about the individual circumstances of the person per day, but what those circumstances say about their prospects in the future. Of course, the credit score sometimes gets it spectacularly wrong. A person who has been in and out of prison might have finally changed their ways and someone that has been a steady payer might have a health crisis that turns them into a defaulter. However, the bank decision is probably right on the vast majority of occasions and that is a level of risk that the lender is able to take on.

The collection of big data can become a sinister exercise (Hair, 2010). This is the case in China where the government is introducing a social credit score that can determine access to house, social services, and even the much-valued exit visa to be able to visit other countries. Chinese citizens are under increased scrutiny under this system. In theory, someone that is late on their rent can be prevented from going on holiday. This level of control can be unnerving to people who live in Western democracies where the growth of big data is viewed with a lot of suspicion (Howells and Wood, 1993). This does not stop them from using credit scoring which is used throughout financial services (Hilbert and Lopez, 2011). Scoring models process a customer's credit history, loan application, customer data, and similar issues in order to rank-order individuals by their likelihood of making future credit payments on time (McFarlane, 2010). Predictive analytics is an area of statistics that deals with extracting information from data and using it to predict trends and behavior patterns (Awang et al., 2013; Helmreich, 2000; Hilbert and Lopez, 2011; Howells and Wood, 1993). The enhancement of predictive web analytics calculates statistical probabilities of future events online (Evans, 2009; Ifinedo, 2016; Little, 2002). Predictive analytics statistical techniques include data modeling (Cappellin and Wink, 2009), ML (Jibril and Abdullah, 2013), Artificial Intelligence (Ifinedo, 2016), deep learning algorithms (Kobie, 2015), and data mining (Helmreich, 2000). Often the unknown event of interest is in the future, but predictive analytics can be applied to any type of unknown whether it is in the past, present, or future (Howells and Wood, 1993). This type of retrospective prediction often happens in the crime investigation process where psychometric information and other relevant scores are used to determine the people most likely to have committed a crime (Helmreich, 2000). For example, it is more likely that someone who committed aggravated assault progressed to armed robbery whereas someone that has never been involved in any kind of violence might still steal but using white-collar crime methodologies. The core of predictive analytics relies on capturing relationships between explanatory variables and the predicted variables from past occurrences (McFarlane, 2010). The analysts then exploit these relationships in order to predict the unknown outcome (Jansen et al., 2008). It is important to note, however, that the accuracy and usability of results will depend greatly on the level of data analysis and the quality of assumptions (Mosher, 2013). The predictive and explanatory theories of data science always emphasize the fallibility of all the output from these models (Hilbert and Lopez, 2011).

Some have argued persuasively that predictive analytics is really predicting a detailed level of granularity (Awang et al., 2013). Therefore, it focuses on generating certain predictive scores or probabilities for a given variable or entity to engage in certain behavior (Hilbert and Lopez, 2011). That is a different approach from one which is used in forecasting. In the case of predictive analysis, the technology that is used is configured in such a way as to learn from the past and present in order to map the future (Jibril and Abdullah, 2013). In future industrial systems, the value of predictive analytics will be to predict and prevent potential issues to achieve near-zero break-down and further be integrated into prescriptive analytics for decision optimization (Davis et al., 2014; Lyytinen et al., 2016; Zhang and Chen, 2015). Furthermore, the converted data can be used for closed-loop product life cycle improvement which is the vision of the Industrial Internet Consortium (Bachman, 2013; Hilbert and Lopez, 2011; Little, 2002). The predictive analytics process has been mapped in order to include certain essential processes as follows:

- **Phase 1 – Define Project:** This phase involves defining the project outcomes (Hilbert and Lopez, 2011), deliverables (Howells and Wood, 1993), scope of the effort (Ifinedo, 2016), business objectives (Hilbert and Lopez, 2011), and identifying the data sets that are going to be used (Rachuri et al., 2010).
- **Phase 2 – Data Collection:** This phase involves data mining for predictive analytics (Hair, 2010). In this phase, the analyst prepares data from multiple sources for analysis (Menke et al., 2007). This provides a complete view of customer interactions (McFarlane, 2010).

- **Phase 3 – Data Analysis:** This phase involves data Analysis is the process of inspecting (Kim and Jeon, 2013), cleaning (Min et al., 2008), and modeling data (Howells and Wood, 1993). The analyst does this with the objective of discovering useful information (Zhang and Chen, 2015), and arriving at conclusion (Sakuramoto, 2005).
- **Phase 4 – Statistics:** This phase involves statistical analysis (Holmes, 2005). This phase therefore enables the analyst to validate the assumptions (Malathy and Kantha, 2013), hypothesize about them (Hamari et al., 2015), and test these hypotheses using standard statistical models (Mieczakowski et al., 2011).
- **Phase 5 – Modeling:** This phase involves predictive modeling (Gilks, 2016). It therefore provides the analyst with the ability to automatically create accurate predictive models about future (Jansen et al., 2008). There are also options to choose the best solution with multi-modal evaluation (Helmreich, 2000).
- **Phase 6 – Deployment:** This phase involves predictive model deployment (Ifinedo, 2016). It therefore provides the analyst with the option to deploy the analytical results into everyday decision-making process (Hilbert and Lopez, 2011). This, in turn, helps to get results (Holmes, 2005), reports (Jansen et al., 2008), and output by automating the decisions based on the modeling (Jansen et al., 2008).
- **Phase 7 – Model Monitoring:** In this final phase, the models are managed and monitored to review the model performance to ensure that it is providing the results expected (Dutse, 2013). There may be changes to the original process based on the new information that is coming in (Hamari et al., 2015). Part of being a dynamic organization is the ability to make changes when appropriate (Mieczakowski et al., 2011).

The predictive and explanatory theories of data science also deal with some of the typologies that are commonly used in industry (Hamari et al., 2015). Generally, the term predictive analytics is used to mean predictive modeling, "scoring" data with predictive models (Holmes, 2005), and forecasting (Ifinedo, 2016). However, people are increasingly using the term to refer to related analytical disciplines, such as descriptive modeling (Hilbert and Lopez, 2011), decision modeling (Menke et al., 2007), and optimization (Lewis, 1996). These disciplines also involve rigorous data analysis (Hilbert and Lopez, 2011). They are also widely used in business for segmentation and decision making (Howells and Wood, 1993), but have different purposes and the statistical techniques underlying them vary (Lewis, 1996). Predictive modeling uses predictive models to analyze the relationship between the specific performance of a unit in a sample and one or more known attributes and features of the unit (Bansal, 2013; Howells and Wood, 1993; Gibson and Brown, 2009). The objective of the model is to assess the likelihood that a similar unit in a different sample will exhibit the specific performance (Bansal, 2013; Malathy and Kantha, 2013; Jansen et al., 2008). This category encompasses models in many areas including marketing (Holmes, 2005), where they seek out subtle data patterns to answer questions about customer performance (Kirchweger et al., 2015). They can also be utilized as a part of different fraud detection models (Bachman, 2013; Hilbert and Lopez, 2011; Noughabi and Arghami, 2011). Predictive models often perform calculations during live transactions (Awang et al., 2013; Min et al., 2009; Sinclaire and Vogus, 2011). For example, they can be used to evaluate the risk or opportunity of a given customer or transaction, in order to guide a decision (Gilks, 2016). With advancements in computing speed, individual agent modeling systems have become capable of simulating human behavior or reactions to given stimuli or scenarios (Menke et al., 2007).

The available sample units with known attributes and known performances are referred to as the "training sample" (Davis et al., 2014). The units in other samples, with known attributes but unknown performances, are referred to as "out of [training] sample" units (Jansen et al., 2008). The out of sample units do not necessarily bear a chronological relation to the training sample units (Kees et al., 2015). The out of sample unit may be from the same time as the training units, from a previous time, or from a future time (Malathy and Kantha, 2013). Descriptive models quantify relationships in data in a way that is often used to classify customers or prospects into groups (Helmreich, 2000). Unlike predictive models that focus on predicting a single customer behavior such as credit risk (Awang et al., 2013; Evans, 2009; Min et al., 2009), descriptive models identify many different relationships between customers or products (Bachman, 2013; Hair, 2010; McFarlane, 2010; Wallace, 2004). Descriptive models do not rank-order customers by their likelihood of taking a particular action the way predictive models do (Bachman, 2013; Evans, 2009; Mieczakowski et al., 2011; Zhang and Chen, 2015). Instead, descriptive models can be used to categorize customers by their product preferences and life stage as a case in point (Hilbert and Lopez, 2011). Descriptive modeling tools can be utilized to develop further models that can simulate a large number of individualized agents and make predictions (Bansal, 2013; Helmreich, 2000; McFarlane, 2010; Ruben and Lievrouw, n.d.).

CHAPTER 4: SUMMARY

This chapter sought to describe some of the statistical applications of data science. The first section showed that the public sector has been able to adapt some of the methods and techniques of data science in order to improve their own performance in terms of reaching as many qualifying service users as possible. The second section showed that data could become a competitive advantage if it is well-sourced, stored, and analyzed in order to create a comprehensive image of the environment within which an organization operates. The third section showed that data engineering practices are designed to ensure that the complex patterns of data mapping are properly identified and then linked to practical decisions within an organization. The fourth section indicated that applied data science is a growing, but complex field within the discipline which will determine how we live our lives in the future. The fifth section indicated that both explanatory and predictive theories help to underpin the activities that are undertaken in the pursuit of using big data in order to make sustainable business decisions. The next section will explore the future of data science.

CHAPTER 5

THE FUTURE OF DATA SCIENCE

CONTENTS

The aim of this chapter is to explore what the future of data science might be like based on what has happened in the past, what is happening now and what is predicted to happen in the future. The first section will highlight the increased reliance on open science as opposed to paid-for services. The second section will consider the co-production and co-consumption that is associated with modern data science.

The third section highlights the importance of data reproducibility as part of the decision-making process. The fourth section will consider how transparency is a necessity when using data science to make decisions.

The fifth section of this chapter will highlight the changing research paradigms in data science within academia. Overall, this chapter will show the kind of issues that organizations and businesses have to consider when they are preparing for a future that is data-driven.

5.1. INCREASED USAGE OF OPEN SCIENCE

When Open Science started as a specialization, many did not take it seriously enough (Awang et al., 2013). At first, it was a group of analysts who were disillusioned with the over-commercialization of the sector (Berker et al., 2006).

Later on, members of the public joined the movement in rejecting the notion that everything has to be paid for in order to be valid (Gibson and Brown, 2009).

There was also some revulsion at the notion that some big corporations could effectively co-opt existing data analytical frameworks for their own commercial benefit (Helmreich, 2000). The principle is that where research is publicly funded, it should be more widely accessible to the community (Noughabi and Arghami, 2011).

The digital formats and the internet, in particular, have made this transition much easier (Mieczakowski et al., 2011).

Even those business entities that might have commercial interests could benefit from having wider access to comprehensive data (Ulloth, 1992). Figure 5.1 highlights the increased usage of open science.

Figure 5.1. Increased usage of open science.

Source: LIBER.

The development of open science arises out of the conflict between the modern commercialization of science and the old traditions of openness in science (Davis et al., 2014). The tools of information and communication technology have a profound impact on scientific enterprise (Hamari et al., 2015). There is recognition of the fact that scientists must be compensated for their innovation (Hamari et al., 2015). At the same time, it must also be acknowledged that a lot of science is financed through public means (Howells and Wood, 1993). That means that the community at large has an interest on how that science is acquired and how it is distributed (Hilbert and Lopez, 2011). Rather than oppressing either side, the new settlement focuses on sharing the benefits (Lyytinen et al., 2016). Open science is the way in which the public can access the results of scientific research (Noughabi and Arghami, 2011). However, there are also certain premium services which have to be paid for so that the scientist secures some level of compensation (Howells and Wood, 1993). This compensation may not always be in terms of monetary rewards, but instead could be the recognition that the person has been widely published in peer-reviewed journals (Min et al., 2008).

Open science serves the additional purpose of promoting scientific research (Dutse, 2013; Holmes, 2005; Min et al., 2008). There are also policy implications for the funders of research as well as institutions of higher learning which have a lot of interest in ensuring that data science is promoted (Hamari et al., 2015). The advent of the internet and other online forums has increased the opportunities to organize and publish research projects (Dutse, 2013). Given the contributions of the online communities, it becomes unconscionable to exclude them from the output that arises out of the research process (Gibson and Brown, 2009). There is a lot of data that is being collected by researchers from members of the public (Hamari et al.,

2015). This goes through a rigorous process of informed consent (Hilbert and Lopez, 2011). However, as a means of giving back to those communities; it makes sense to have some provisions for open data science for those members of the community that are unable or unwilling to pay for access (Lyytinen et al., 2016). Where the scientific community is unable to or unwilling to share, information freely; there is evidence that there are online communities that are able to access and distribute information using alternative and free sources (Hamari et al., 2015). Perhaps the best example of this is Wikipedia whose model relies on editing and contributions by members of the public (Miller, 2014). The website has rapidly become a significant reference point to the extent that some politicians who wish to raise their profile attempt to corrupt the process of updates in order to erase negative coverage about their lives (Berker et al., 2006; Ifinedo, 2016; Wallace, 2004).

Information and communication technologies allow for the collection and sharing of large amounts of data (Hair, 2010). This information is critical for the various scientific experiments that take place on a regular basis (Holmes, 2005). Therefore, science has become increasingly data-driven (Jansen et al., 2008). This creates a moral obligation to support open data science because the entities that contribute to the scientific process have a right to share in the results of the study (Jansen et al., 2008). The principle or reciprocity and mutuality can go a long way to reassure communities that are concerned that scientists are exploiting them in order to produce commercially available data without leaving something behind or giving back to the various communities that have helped them (Sinclaire and Vogus, 2011). Online archives are a very important tool for storing big data and also increasing its accessibility to a wide audience (Gibson and Brown, 2009). For some data scientists, this is an ideal situation that must be actively cultivated in order to reach the widest audience possible (Jibril and Abdullah, 2013). Besides, the speed of the data transactions also means that decision-making is made easier by virtue of the fact that the people that need information can get it as and when they need it without having to engage in significant alternative research (Hilbert and Lopez, 2011). Paul David coined the term "open science" in 2003 (Dutse, 2013). This term was understood to encompass all those scientific goods that were generated by the public sector or on behalf of the public sector (Ifinedo, 2016). This phenomenon was always perceived to be in possible to the extension of stringent and commercialized intellectual property rights to data science (Jansen et al., 2008). Some economists have argued that all scientific knowledge that is generated through public research must then be handled as if it were a public good (Engelberger, 1982). This can be demonstrated in practical terms when everyone that wishes to engage with this knowledge is capable of accessing it (Hamari et al., 2015).

5.2. CO-PRODUCTION AND CO-CONSUMPTION OF DATA SCIENCE

The open science that has been described in the previous section is rationalized through the co-production and co-consumption of data science (Ellison, 2004; Evans, 2009; Min et al., 2009). Co-production is a form of collaborative research which extends the participation of various stakeholders (Evans, 2009). The reasons for this may include the need to facilitate research and extend its impact much wider than previous modalities (Hamari et al., 2015). There are many arguments in favor of coproduction which have been explored in existing research (Evans, 2009). Moreover, such research has indicated some of the optimal means of creating co-productive research units including practical considerations such as cost-sharing (Holmes, 2005). Before exploring the benefits of co-production and co-consumption of data science, it is important to understand the situations where they are appropriate modalities or where they would be inappropriate (Ifinedo, 2016). This is also part of the decision-making process (Noughabi and Arghami, 2011). Figure 5.2 shows that big data is very transactional and that provides a rationale for engaging in transactional research in order to access it (Holmes, 2005).

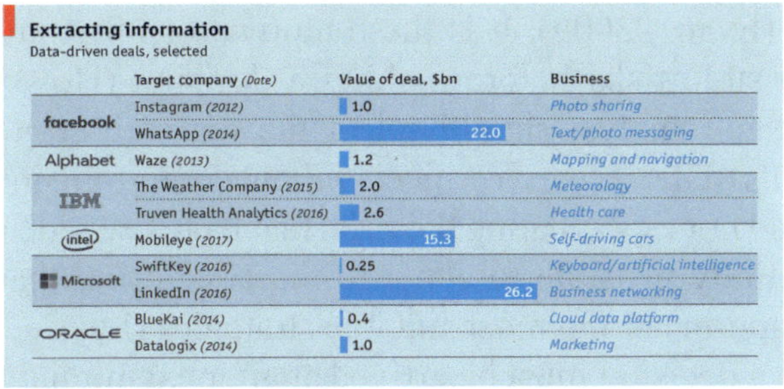

Figure 5.2. Data-driven transactions.

Source: The Economist.

The multiplicity of justifications for coproduction has not meant that the debate about its efficacy is fully resolved (Hair, 2010). Some of the issues that are contentious include the rationale for coproduction (Holmes, 2005), its impact (Min et al., 2008), and the most optimal approaches to work with (Rachuri et al., 2010). Nevertheless, co-production, and co-consumption of data science can help to change populations (Menke et al., 2007), institutions (Schute, 2013), and paradigms (Spiekermann et al., 2010). There are certain risks that are associated with coproduction which must be taken into consideration when setting up a research modality (Gilks, 2016). For example, the various actors within the team may disagree about the direction of the research and even the findings (Howells and Wood, 1993). One of the mitigations that can be put in place is to have a hierarchy of researchers so that someone is able to make decisions that are binding on the rest of the team (McFarlane, 2010). There are vie costs that are mainly associated with co-production in research (Helmreich, 2000). These costs can affect the research (Holmes, 2005), the process (Malathy and Kantha, 2013), the professionals involved (Menke et al., 2007), other stakeholders (Mieczakowski et al., 2011), scholarship in general (Mosher, 2013), and the community at large (Bachman, 2013). The problem is that existing literature rarely references these costs, but instead emphasizes the benefits of coproduction (Gibson and Brown, 2009). Indeed, there is a trend in which greater inclusion in the research process is recommended (Gibson and Brown, 2009). Despite the costs and risks that are associated with coproduction in data science; there will be tools, methods, and approaches that can be used to mitigate them (Lewis, 1996).

There are specific motivations that might inspire an analyst to engage in co-production (Evans, 2009). It is these motivations that must be weighed against the possible risks before making a decision (Howells and Wood, 1993). Ideally, the analyst should select the research strategies that best achieve the goals of the research project without compromising on the quality of the output and not exacerbating the risks that have been identified (Ifinedo, 2016). It is entirely possible to come to the conclusion that coproduction and co-consumption are the most suitable strategies (Lyytinen et al., 2016). However, such a decision must be arrived at after examining all the relevant facts (Ulloth, 1992). Although there is a presumption that coproduction is the ideal, some caution is called for in light of some of the costs that are associated with this type of research modality (Helmreich, 2000). The analyst should adopt an open mind so that they are able to consider some of the options that may not be immediately obvious, but which can actually deliver the goals that they are aiming for (Howells and Wood, 1993). Reflective research techniques can ensure that these considerations are given the importance that they deserve rather than going ahead with specific actions that appear to be initially advantageous but which may impact on the utility of the research process (Lyytinen et al., 2016).

It is worthwhile to consider the unique ways in which coproduced data science has been able to change the world or the environment within which an organization is operating (Hair, 2010). This can be gleaned from pre-existing research projects which might be under obligation to explain the modalities that they relied on (Mosher, 2013). A cost and benefits analysis may be called for when considering this option, particularly if the analyst of the commissioning organization has other alternatives that it can explore (Gibson and Brown, 2009). The team may seek preliminary advice from experts and people who are experienced in working with coproduced data science so that they can give them insights into some of the issues that they are likely to encounter if they go ahead with the planned project (Jansen et al., 2008). The agreement on the various protocols might include some type of "translation" so that the highly technical information is transformed into formats that are easily understandable (Min et al., 2009). It is worthwhile to always consider the fact that often policymakers are not particularly adept at the more technical aspects of data science (Hamari et al., 2015). Some researchers are even questioning the premises under which decisions are made about the ontology and epistemology of data science (Engelberger, 1982). Some of the assumptions that have held sway for a long time are no longer applicable given the changes that are taking place in the business world (Hamari et al., 2015). There has to be an effort to convert technical data science into practical information that is easily accessible to the people that are supposed to use it, regardless of the position which they occupy within the organization (Holmes, 2005). The organization may also develop research skills and data management skills that can be utilized in other aspects of the organization's operations (Bansal, 2013; Holmes, 2005; Ruben and Lievrouw, n.d.).

5.3. BETTER REPRODUCIBILITY OF DATA SCIENCE

One of the performance measures that are currently being explored by business is that of achieving improved reproducibility of data science output (Carlson, 1995). In order for data science to be truly valuable, it should be reproducible, reusable, and robust (Hair, 2010). This is also the foundation of deep reinforcement learning from that data (Howells and Wood, 1993). This is a concern that goes beyond traditional data science (Jansen et al., 2008). For example, a 2016 survey by Nature magazine found that the vast majority of scientific fields are facing a reproducibility crisis (Kees et al., 2015). The irony is that despite the increasing amount of data that is on the market today, one of the issues that lead to poor reproducibility is insufficient statistical knowledge (Chesbrough, 2005). That could be a consequence of poor data collection techniques or the difficulty in obtaining high-quality data (Hilbert and Lopez, 2011). There are specific disciplines that face specific problems (Evans, 2009). For example, medicine, and biology have a high number of respondents (approximately 60%) who are not trained in statistics. Of course, the assumption is that data analysts would be definition have good statistical training which somewhat mitigates this problem (Min et al., 2008). Figure 5.3 shows a generic data management framework for academic research which is somewhat more stringent than the generalized protocols for data science (Evans, 2009; Mosher, 2013; Zhang and Chen, 2015; van Deursen et al., 2014).

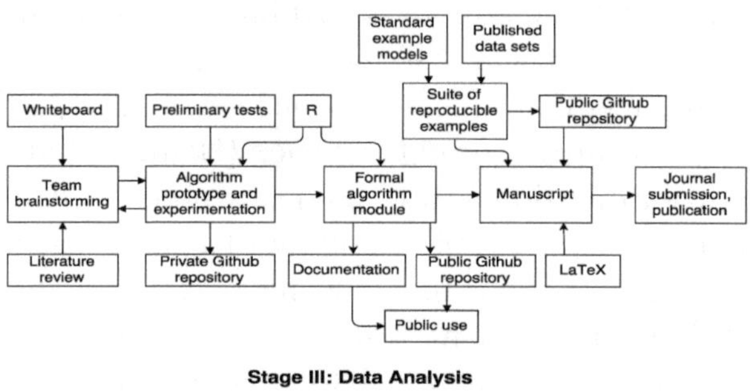

Figure 5.3. Data analytical frameworks.

Source: eScience.

Despite the emphasis on deterministic modeling and statistics, machine learning (ML) practitioners are still reporting significant problems of reproducibility (Hamari et al., 2015). The structural and organizational barriers to getting good data have not yet been fully surmounted (Howells and Wood, 1993). Existing literature has shown that these barriers would require a complete overhaul of the entire system before they could be successful transformed (Jibril and Abdullah, 2013). The transitional period would place considerable pressures and responsibility on the practitioners (Sin, 2016). According to Chris Drummond, many of the discourses on reproducibility end up being discussion on replicability (Jansen et al., 2008). In these simplest terms, this means that the same methodologies produce the same results or data (Kees et al., 2015). Whereas computational models may tolerate inaccuracies, the real world is less forgiving (Jibril and Abdullah, 2013). For example, open-source code produces the same data and results when run across various applications (Kees et al., 2015). Meanwhile, experiments are rarely completely replicable since there are many intangibles that are a one-off for that experiment (Hilbert and Lopez, 2011). The implication is that each experiment is actually a unique experience in a unique situation (Howells and Wood, 1993). We make allowances because the variances are not so high as to render the experiment unrepeatable (Hilbert and Lopez, 2011). The fidelity of experimental replication differs between laboratory and computational disciplines (Kirchweger et al., 2015). The fidelity of computational replication is generally expected to be incredibly high (Bansal, 2013; Helmreich, 2000; Min et al., 2009). If another researcher applies the same code to the same data, it would be expected that any deterministic algorithm would produce the same or very similar results (Cappellin and Wink, 2009; Ellison, 2004; van Deursen et al., 2014). Essentially, most open source projects meet this replicability requirement and that is one of their key characteristics (Bachman, 2013; Ellison, 2004; Sin, 2016). The implication is that merely stopping at this level of experimental reproduction is likely to be trivial for most of the meaningful research in the field (Bachman, 2013; Gilks, 2016; Little, 2002; Stone et al., 2015).

The triviality notwithstanding, this sort of exercise may still be critically important to serve as a positive control for other practitioners rolling out a new tool or algorithm (Bachman, 2013; Gibson and Brown, 2009; Little, 2002). According to Drummond, reproducibility involves more experimental variation (Bansal, 2013; Howells and Wood, 1993; McFarlane, 2010). It is almost natural to think of experimental reproduction as an activity that exists on a continuum from near-perfect similarity to complete dissimilarity (Bachman, 2013; Helmreich, 2000; Min et al., 2008). On the high-fidelity end of the scale, we have a forked project re-executed with no changes (Engelberger, 1982; Howells and Wood, 1993; Mieczakowski et al., 2011). Low fidelity produces a result that is nothing like the anticipated outcome, given the fact that there are so many substitutions that the second experiment cannot be considered to be a replication of the first (Hair, 2010; Jibril and Abdullah, 2013; Zhang and Chen, 2015). Hence, experimental replication in a laboratory experiment looks more like reproduction in a computational experiment (Hamari et al., 2015; Noughabi and Arghami, 2011).

The analysis above shows that an effective reproduction rating for a data experiment is a compromise between complete replication and complete irrelevance (Dutse, 2013). We ought to pay attention to the issue of reproducibility because it is the foundation of data science and its methodologies (Helmreich, 2000). Single occurrences that cannot be reproduced serve very little purpose in science because they cannot be subjected to experimentation, testing, hypothesis, and conclusion (McFarlane, 2010). They are what they are and cannot be changed thereafter (Min et al., 2008). Scientists are justifiably suspicious of people that claim to be the only ones that can produce certain results and yet demand resources for that one-off research that they have done (Hair, 2010). It is why academic journals pay particular attention to the methodologies so that other scientists might be able to conduct the same or similar experiments in order to ascertain whether or not they achieve the same results (Hamari et al., 2015). Single and unconfirmed anecdotes cannot be the foundation of data science and those organizations that rely on such experiments are soon going to find that they have made the wrong choice (Evans, 2009). It is important to note that not all instances of irreproducibility are malicious or otherwise intentional (Holmes, 2005). Sometimes there are unique circumstances that produce outlier results. Data science is founded on the notion that methods can be replicated by others or reproduced in order to achieve the same or similar results (Jibril and Abdullah, 2013). Above all, data science should always aim to unmask and resolve problems rather than hiding them away behind technical details and jargon (Gilks, 2016).

5.4. TRANSPARENCY IN THE PRODUCTION AND USE OF DATA SCIENCE

Whereas technical prowess helps to achieve impressive data science, there is a need to ensure some level of transparency during the production process (Gilks, 2016). It is through this transparency that it is possible for others to challenge the assertions and assumptions and emanate from research (Holmes, 2005). The industrial complex that dominates contemporary life calls for the digitization of everything including factories in order to make savings and increase profits (Jansen et al., 2008). The move towards digitization is not really a choice because those organizations that do not digitize are bound to fail in a modern economy since they miss out on the competitive advantages of big data (Jansen et al., 2008). In order to make sound decisions that are data-driven, it is important to have real-time visibility in the operations of the structures and infrastructure that underpins an organization's performance (McFarlane, 2010). Managers are no longer spending as much time as they should on a strategy based on empirically tested predictions (Evans, 2009). Instead, managers have adopted a reactive stance to the data deluge which can closely resemble the 24-hour news cycle (Hair, 2010; Jansen et al., 2008). They do not spend the necessary effort identifying the root causes of business problems or engaging in deep retrospection. Instead, they are firefighting with everything that they have (Holmes, 2005). Figure 5.4 demonstrates a tentative transparency framework that can allow for reflection and considered business decision making after reviewing big data.

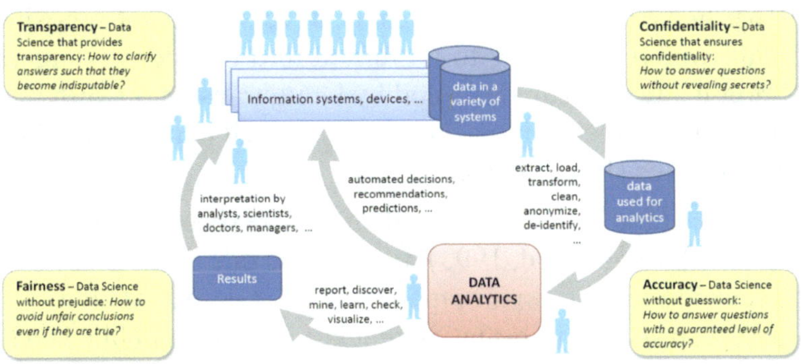

Figure 5.4. Transparency frameworks for data science.

Source: Redasci.

Some firms have sought solutions to the problems they face by trying to consolidate all the data that they need into a single or a number of repositories that can be easily accessed by their staff. This is achieved through the use of software applications such as enterprise resource planning (ERP) and business intelligence (BI). This is possible if all the hardware and software in factories are connected and communicate (exchange data) with each other. In that sense, the way in which information is processed and received is determined by the existing structures that are designed to handle that information (Evans, 2009). When advocating for transparency in the production and use of data science, it is important to incorporate this into an overall strategy that is focused on ensuring that decisions are backed by data and that they are proactive in their stance (Gilks, 2016). Using big data in this way can allow the firm to develop a deep understanding of all its constituent parts which are cost centers at the end of the day (Ellison, 2004). This approach is justified by the direct impact that it has on the bottom line (Hamari et al., 2015). BI drives performance in the modern factory or organization (Gibson and Brown, 2009). However, that intelligence is an important asset which the company may want to protect from third parties that are direct competitors or are otherwise engaged in activities that are harmful to the organization (Jibril and Abdullah, 2013). This is where the need for transparency in the production and use of data science can be question (Little, 2002). Some might argue that the need to protect BI is so strong that it supersedes all other concerns (Malathy and Kantha, 2013).

The nature of BI may be such that it is not very different from what the company usually reviews prior to the introduction of data science (Kees et al., 2015). For example, a factory may have significant repositories of machine data which is waiting to be turned into decisions (Hamari et al., 2015). Through advanced analysis and by being transparent with regards to the decision-makers, the business data can be turned into specific directional goals and objectives (Hair, 2010).

Therefore, BI (and by implications data science) becomes a core aspect of planning (Howells and Wood, 1993). Planning is one of the most significant arenas of decision-making in an organization (Mieczakowski et al., 2011). The company can sue the information that is acquired from big data to make investment decisions that may, for example, ensure that the most productive departments are fully funded while those that are less productive are scaled down or eventually removed (Hair, 2010; Little, 2002). This is a rationale that is backed by science (Ifinedo, 2016). Explaining the rationale to workers might mitigate the risk of suffering a backlash from the affected employees who may be resistant to change (Hamari et al., 2015). This is a case where transparency in the production and use of data science can help to break down barriers to better performance in organization context (Gibson and Brown, 2009). There are five ways in which you can enhance transparency in the production and use of data science:

1. **Edge Connectivity:** It is advisable to build a certain level of edge connectivity using certain tools including OPC-UA, Modbus, Ethernet, and Profinet (Gibson and Brown, 2009). This allows the edge devices to access a common cloud network much more easily and faster (Jibril and Abdullah, 2013). Your preference should be geared towards intelligent connectivity which allows for alerts, decisions, notifications, and light analytics (Ifinedo, 2016).
2. **Manufacturing Data Lakes:** A manufacturing data lake allows you to curate data from various machines using some pre-existing DCS and SCADA systems (Davis et al., 2014). You can also

utilize enterprise data historians and manufacturing execution systems (MES) which are then placed into a single repository (Hair, 2010; Jibril and Abdullah, 2013). In effect, you have all the information that you need in one place which can be convenient for decision-making (Jansen et al., 2008).

3. **Key Performance Indicators (KPIs):** You must make an effort to track your KPIs which will tell you about your operational performance (Helmreich, 2000). These indicators must be specific, measurable, achievable, realistic, and time-bard (Jibril and Abdullah, 2013). You will specifically be interested in the performance that relates to overall equipment effectiveness (OEE), capacity utilization (CU), and machine downtime analysis (Sin, 2016).
4. **Business Intelligence (BI):** It is important to leverage BI by constructing internal and external systems that can develop specific benchmarks (Carr, 2010). In turn, this allows you to create customized reports which can tell you about the state of your KPIs (Chiu et al., 2016). This information can then be easily shared with the rest of the team. Analytics can tell you what is going to happen and this can become a key resource for your staff (Gilks, 2016). By establishing correlation models, you are able to utilize notifications and alerts in order to facilitate all the decision-makers in the organization (Little, 2002).
5. **Integration:** It is an important facet of transparency (Hamari et al., 2015). Business or manufacturing intelligence makes an impact when shared with decision-makers (Gibson and Brown, 2009). Conversely, it has a very limited impact if it is exclusive accessible to a few people that do not make decisions (Kees et al., 2015). Through integration, you will be able to link your machine data to the enterprise system data (Kobie, 2015). This in effect means that you are integrating your operational technology data with the enterprise systems which may include a customer relationship management (CRM) model, enterprise planning, scheduling, and human resource management systems (Hamari et al., 2015). Always ask for customized adapters if available because they give you the freedom to configure your reports as best as can be (Menke et al., 2007).

By connecting machines and collecting data, organizations can gain real-time visibility into factory operations (Awang et al., 2013; Helmreich,

2000; Jibril and Abdullah, 2013; Trottier, 2014). With real-time time-series data, they build a great manufacturing data lake (Abu-Saifan, 2012; Bansal, 2013; Chesbrough, 2005; Holmes, 2005). Using big data and ML, manufacturers can build a predictive model and derive the real benefits from digitalization (Awang et al., 2013; Chesbrough, 2005; Kim and Jeon, 2013; Sin, 2016). These include optimized energy use and a better read of how your organization is using its resources (Awang et al., 2013; Engelberger, 1982; Hilbert and Lopez, 2011; Ruben and Lievrouw, n.d.). The resultant improvements can transform the organization in terms of the control that it has over its bottom line. Of course, transparency also comes with risks and the organization has to do a risk assessment to identify this. The next step is to develop contingency and mitigation plans that ensure the appropriate use of data science.

5.5. CHANGING RESEARCH PARADIGMS IN ACADEMIA

Given the transformations that data science is undergoing, the academics that facilitate its development are also changing their research paradigms (Boase, 2008). These shifts reflect the new priorities of the higher institutions of learning as well as the realities of operating within a commercialized environment where costs must be justified by tangible results (Davis et al., 2014). The paradigm shift will affect how data is collected, treated, and shared (Jibril and Abdullah, 2013).

Indeed, the performance indicators for these higher institutions of learning will reflect the changing research paradigms in academia with regard to data science (Lyytinen et al., 2016). Nevertheless, the transition period is not without its challenges and these institutions are grappling with requirements that severely test their resource base and technical competencies (Evans, 2009).

A case in point is how traditional surveys are becoming unsustainably expensive and have to be replaced by online tools; not least, because they have declining response rates (Mieczakowski et al., 2011). Nobody wants to be filling out endless forms in order to suit the data needs of a researcher (Kirchweger et al., 2015). Figure 5.5 highlights some of the key issues that should be of concern to a data scientist in this era.

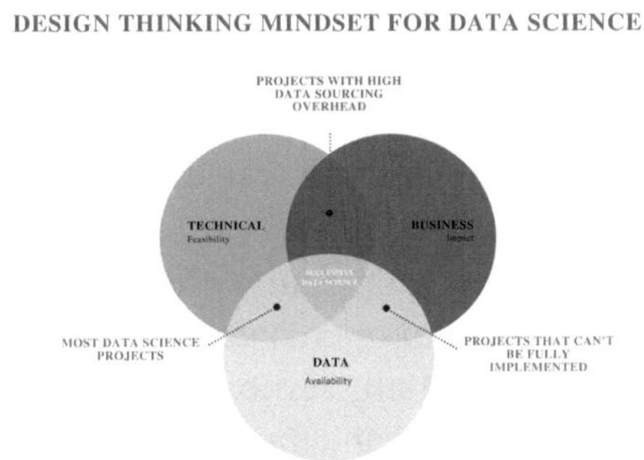

Figure 5.5. Transforming the research paradigms of data science.

Source: Towards Data Science.

Moreover, the exclusivity that academia enjoyed in terms of handling data science projects is being challenged by the notion of co-production and co-consumption (Carlson, 1995). Some organizations have already facilitated their own in-house research teams which produce data which is bespoke to them (Dutse, 2013). In that case, the academics that are doing additional research might begin to lose their relevance to these private corporations that have found alternatives (Malathy and Kantha, 2013). The working methods of the researchers also need an overhaul in order to meet the demands for real-time data (Hair, 2010). For example, it is no longer acceptable that there are months between when the time that data is collected and when a final report of findings is published (Helmreich, 2000). By this time, there is a very high risk that the organizations that commissioned the data will have lost interest (Kobie, 2015). Part of this delay is due to the stringent quality checks that go into ensuring that peer-reviewed content is acceptable (Chiu et al., 2016). However, the reality is that the demands from the organizations that rely on data science are just not in keeping with the traditional ways of working (Hair, 2010; Jansen et al., 2008). That is why some organizations have started setting data traps and filters that weed in the information that is required for their own decision-making in real-time (Lyytinen et al., 2016). This allows them to make decisions immediately and when required, rather than waiting for an academic to complete their research cycle (Mosher, 2013).

Part of the changing research paradigms in academia with regards to data science is the quest for new tools that can facilitate the requirements to manipulate, extract, and analyze data in a different way (Evans, 2009). This difference is primarily in terms of speed (Lyytinen et al., 2016). We live in a disposable information economy where information must be acquired quickly in real-time and then discarded once it has been used for decision-making because it is constantly being replaced by new information (Mieczakowski et al., 2011). Users are making demands on researchers in terms of requirement a host of various repositories for distributing statistical data (Chiu et al., 2016). Researchers must additionally deal with the thin line between private and public repositories in an era of open data science (Kobie, 2015; Rachuri et al., 2010). The internet has virtually removed the transnational boundaries that used to categories data sources and data consumption units (Davis et al., 2014). More developing countries are entering the loop of big data and they have different needs from the established developed economies (McFarlane, 2010). Members of the public are generally more interested in information as it relates to their needs (Gibson and Brown, 2009). They demand that complex statistical measures are broken down into chunks that are easily understandable and relatable to their own experiences (Mieczakowski et al., 2011). Academics that produce data science which is opaque and inaccessible may find that it is not consumed as much as that data science that is deliberately accessible (Miller, 2014). This has led to concerns about the temptations of racing to the bottom in order to attract more users of big data (Bansal, 2013).

All this has led to the urgent need to have new methodologies in order to deal with the data requirements of a consumer that is very different from the people that treated old fashioned research as gospel truth (Gibson and Brown, 2009). This is not necessarily a bad thing for the changing research paradigms in academia with regards to data science (Helmreich, 2000). First of all, analysts are beginning to critically evaluate their own responses and approaches to their work in order to ensure that they meet with the expectations of their clients (Mieczakowski et al., 2011). A case in point is how the standards and taxonomies for data collection and analysis are expected to change over the coming decades (Lyytinen et al., 2016). Similar changes happened between 2000 and 2010 when the OECD's National Experts on Science and Technology Indicators revised the Frascati Manual (OECD, 2002) and the Oslo Manual (OECD-Eurostat, 2005) on a rolling basis (Hamari et al., 2015). The group worked on priority themes and to build a better bridge between the two manuals. The North American Industry Classification System (NAICS) codes and the Standard Occupational Codes also underwent some revision (Berker et al., 2006). These are the types of changes that demonstrate the fact that data science is not a static concern but one which is dynamically changing in response to the demands from consumers (McFarlane, 2010). The rapid changes that are taking place must be incorporated into pre-existing frameworks for data management and data collection (Holmes, 2005). The World Wide Web, in particular, has been transformational in enabling new forecasting and data collection methods that yield useful insights in almost real-time (Kees et al., 2015). These tools provide information much more rapidly than is possible with traditional surveys, which entail up to multiple-year lags (Mosher, 2013).

CHAPTER 5: SUMMARY

This chapter sought to highlight some of the changes that are taking place in data science. The first section showed that the community at large has responded to the exploitative trends in commercialized data distribution by fully embracing the idea of open science. The second section showed that part of this response is to adopt co-production and co-consumption of big data regardless of some of the practical and philosophical limitations of this approach. In the third section, we saw that one of the key quality issues facing data science is that of reproducibility and that the best ways of resolving this problem is to increase the quality of data collection and processing. The fourth section showed that transparency in data science can improve decision making as well as the relationships among the different people that rely on that data. The fifth section of this chapter showed that academia is changing its research paradigms in response to the rapid changes that are taking place in the production and consumption of data science. This chapter has shown that far from remaining a technical concern for industry insiders, data science remains very relevant to all types of organizations and will continue to be so in the foreseeable future. The next chapter considers the curriculum requirements for data science.

CHAPTER 6

THE DATA SCIENCE CURRICULUM

CONTENTS

A new curriculum is required in order to educate the data scientists of the future, a future that the previous chapter has described as being very challenging and requiring the most dynamic competencies. The first section of this chapter will look at the use of advanced probability and statistical techniques. The second section will look at training for working software packages that dominate data science today including Python and Microsoft Excel. In the third section, we will consider the use and development of social statistics as well as their relationship to social enterprise. The fourth section in this chapter considers the requirements for computational competence among business leaders. The fifth section will highlight the language requirements of data science. The overall objective of this chapter is to provide the reader with an overview of some of the critical areas that might be included in a curriculum development program for data science.

6.1. ADVANCED PROBABILITY AND STATISTICAL TECHNIQUES

Data science must necessarily rely on advanced probability and statistical techniques for producing the kind of output that will meet the emerging need in the business community (Chiu et al., 2016; Gibson and Brown, 2009). The popularity of data science in academia has not completely erased the negative stereotypes surrounding the discipline as being rather opaque and difficult to relate to (Evans, 2009). The fundamental concepts of data science have not yet been fully defined and that means that some potential applicants are not really sure about what they are studying or how useful it will be for their careers (Holmes, 2005). The advanced probability and statistical techniques for data science which are at the center of the course may be covered by other fields such as mathematics and even statistics (Menke et al., 2007). At other times, they are incorporated as a sub-topic for research training in traditional disciplines such as education, social work, medicine, engineering, architecture, and law (Ulloth, 1992). Typically, the starting point will involve developing a solid understanding of how algorithms work (Kirchweger et al., 2015). This can start in the first year of study and continue throughout (Kirchweger et al., 2015). The emphasis is on using knowledge about algorithms to reformulate data in ways that are conducive to decision making (Trottier, 2014). By making this type of emphasis, it is possible to create a new market for the output that higher learning students create (Trottier, 2014). The information that they eventually produce will be useful to business (Zhang and Chen, 2015).

One thing that must always be clear to new students is the fact that the advanced probability and statistical techniques for data science are a necessary aspect of the course (Gilks, 2016). They cannot be escaped and perhaps an aptitude test might be appropriate in order to identify those that are not able to deal with the rigors of the course, regardless of their enthusiasm for it (Min et al., 2009). This is in additional to the other professional skills that can turn the student into a practitioner that is capable of handling even the biggest data projects on the market (Kees et al., 2015). Probability is the likelihood of an event or sets of event happening (Helmreich, 2000). In data science, probability is the predictive mechanism because it turns raw data in the present and the past into fairly accurate approximations of what is likely to happen in the future (Mieczakowski et al., 2011). When entrepreneurs are talking about their intuition or sixth sense, they are actually referencing probability but in a non-scientific way (Howells and Wood, 1993). The probabilities of data science are carefully calculated so that the prediction that they support is fairly accurate given the circumstances (Jansen et al., 2008). The world which businesses and organizations have to navigate is unpredictable and that can lead to uncertainty (Hair, 2010). In order to remove this uncertainty when making decisions, we use the informed probabilities of data science (Min et al., 2009). In Figure 6.1, we see how the predictive analytics approach requires very extensive assumptions about the actions and reactions of the major players within the environment (Lewis, 1996).

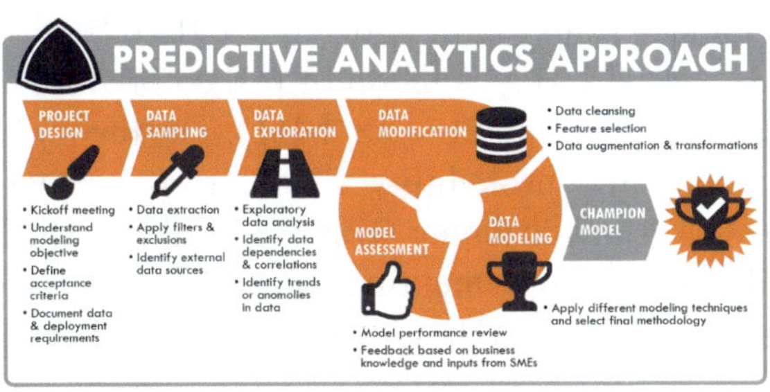

Figure 6.1. Predictive analytics approaches in data science.

Source: Pinnacle Solutions.

Data science is therefore an attempt to control the environment by trying to work out how the variables of the environment will react to specific situations (Engelberger, 1982). Randomness and uncertainty are imperative in the world (Helmreich, 2000). Therefore, it can prove to be immensely helpful to understand and know the chances of various events (Lyytinen et al., 2016). Learning of probability helps you in making informed decisions about likelihood of events, based on a pattern of collected data (Berker et al., 2006; Min et al., 2008; Zhang and Chen, 2015). In the context of data science, statistical inferences are often used to analyze or predict trends from data, and these inferences use probability distributions of data (Kees et al., 2015). Thus, your efficacy of working on data science problems depends on probability and its applications to a good extent (Gilks, 2016). This is a principle that will be operationalized throughout the career of the data analyst (Hair, 2010). They are constantly looking for ways to perfect their prediction based on an intimate understanding of all probabilities and how other external factors affect them (Lewis, 1996). Conditional probability is a natural occurrence that affects experiments where the outcomes of one initial trial are likely to affect the outcomes of additional trials (Ellison, 2004). The analyst will consider all the evidence, assertions, presumptions, and assumptions that underpin all the probabilities in the chain (Little, 2002). Therefore, if the probability of the event modifies when the first event is taken into consideration, it can be said that the probability of the second event is dependent on the occurrence of the first event (Gibson and Brown, 2009; Hilbert and Lopez, 2011). A number of data science techniques depend on Bayes theorem (Carr, 2010; Miller, 2014; Zhang and Chen, 2015). This is a formula that demonstrates the probability of an event depending on the prior knowledge about the conditions that might be associated with the event (Bansal, 2013; Holmes, 2005; Kim and Jeon, 2013). Reverse probabilities can be found out using the Bayes theorem if the conditional probability is known to the analyst (Hilbert and Lopez, 2011). The theory helps to develop a predictive model of the probability of the response variable of some class, given a fresh set of attributes (Hamari et al., 2015; Lyytinen et al., 2016; Stone et al., 2015). Implementing of code interconnect the knowing possibilities (Davis et al., 2014; Hair, 2010; Mieczakowski et al., 2011).

The next challenge is then deciding whether the advanced probability and statistical techniques for data science are undertaken as a technical exercise within the field or whether they are applied at all times to the business environment (Gilks, 2016). In all probability (pun fully intended), the vast majority of entrepreneurs are not familiar with advanced statistic methods such as regression analysis or even the intricacies of predictive probabilities (Menke et al., 2007). They may have a sense that the future can be predicted based on the present and past (Malathy and Kantha, 2013). However, that is different from having advanced technical skills and knowledge (Spiekermann et al., 2010). The curriculum must therefore respond to these decision-makers by training analysts to make the transition to and from the technical aspects towards the practicalities of data science and what they mean for the organization (Ulloth, 1992). The competencies of the data scientist are not so much related to the complexity of their analysis as it is related to their ability to communicate effectively with their clients (Ulloth, 1992; van Nederpelt and Daas, 2012). This aspect of communicating with the client is one of the distinguishing features of those analysts that are able to forge a career in whatever industry they are placed in (Min et al., 2009).

6.2. SOFTWARE PACKAGES SUCH AS MICROSOFT EXCEL AND PYTHON

The adoption of software packages for data science is widespread in order to make it easier to handle some of the more complicated pathways for making decisions based on big data (Cappellin and Wink, 2009; Lewis, 1996; Zhang and Chen, 2015). The data analysts that are being trained to support businesses today must learn how to use these packages so that they are capable of undertaking their work in a manner that is satisfactory (Lyytinen et al., 2016). The software packages that are highly ranked and regularly used keep changing (Holmes, 2005). Therefore, it is imperative to have a dynamic curriculum that can accommodate these changes (Hilbert and Lopez, 2011). For example, Sisense was voted the best data analytics software in 2019. Of course, these decisions are inherently subjective (Lyytinen et al., 2016). However, the core functionalities that are useful to the data analyst are fairly objective including the ability to aggregate large data sets, visualize results so that they are accessible for the decisions maker, and undertake fast analysis in real-time (Sakuramoto, 2005). There may be various specializations even when working with the software packages (see Figure 6.2).

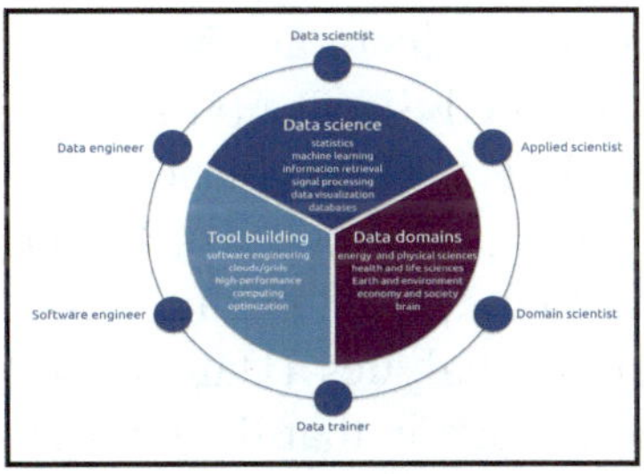

Figure 6.2. Professional specializations in data science.

Source: Data Science Paris.

When commissioning software packages for data science, it might be appropriate to consider other performance attributes that will help decision-makers (Dutse, 2013). A case in point is the use of scalable architecture which is ideal for handling a wide range of data sets with large volume (Holmes, 2005). This facility can be operationalized in small, medium, and large businesses (Min et al., 2008). The digital age has meant that it is much easier to share information and access it in order to support decision making (Dutse, 2013). The fact that information is readily available enables managers to make smarter decisions (Hilbert and Lopez, 2011). For an organization, there might be logistical problems when trying to collect and process various data sets (Hamari et al., 2015). The requirements for human resource input might take away from the other operations within the company (Kobie, 2015). Using advanced algorithms and artificial intelligence allows you to transform the raw data into insights that are valuable for purposes of decision making. Technology means that these insights can be accessed at a click of a button (Hair, 2010; Lyytinen et al., 2016). There are plenty of factors involved in finding the right analytics tool for a particular business (Hair, 2010). From checking its performance and learning the strong suit of the platform to figuring out how well it plays with other systems and identifying its stated performance; the whole research process can be overwhelming to businesses that have a lot of other operations concerns that they must address (Bansal, 2013; Helmreich, 2000; Min et al., 2009). One of the solutions is to visit professional websites that compare and contrast the various products on the market (Miller, 2014). These sites are normally handled by people who have a pre-existing interest in big data and data science generally (Sakuramoto, 2005). They can take a critical stance when new software is presented on the market and give you tips on the kind of products that might be suitable for your business (Sobh and Perry, 2006).

Currently, a number of products are being promoted on the internet pages including the highly-rated Sisense (Gilks, 2016). Others are Looker, Periscope Data, Zoho Analytics, Yellowfin, Domo, Qlik Sense, GoodData, Birst, IBM Analytics, Cognos, IBM Watson, MATLAB, Google Analytics, Apache Hadoop, Apache Spark, SAP business intelligence (BI) Platform, Minitab, Stata, and Visitor Analytics (Hamari et al., 2015). The biggest benefit that these programs have is that they allow you to handle complex data in a fast and efficient way (Cappellin and Wink, 2009). Using scalable systems will mean that you do not have to worry about the daily growth of data volumes since the system already has capacity to expand according to your needs (Helmreich, 2000). Kaggle organized a survey on data science and machine learning (ML) among over 15,000 data industry professionals who were recruited in over 170 countries (Chiu et al., 2016). The survey revealed a number of challenges that should be mitigated during the process of commissioning data science in order to avoid preventable bottlenecks (Howells and Wood, 1993). The biggest challenge that the respondents identified is data that is filled with noise and therefore has to be cleaned up thoroughly before it can be processed (Carlson, 1995). The next problem was related to the lack of suitably qualified professionals that are able to interpret the data and advise those that are making decisions (Gibson and Brown, 2009). In the third place was company politics which prevented the analysts from doing their work properly since a lot of their effort was wasted on trying to reconcile warring factions within the organization (Ifinedo, 2016). The next problem related to poor research briefs that did not set out the exact questions for which the client wanted answers (Jibril and Abdullah, 2013). As a consequence, the analysis that was done failed to meet the expectation of clients and that may lead to the termination of the contract before the entire report has been presented (Hair, 2010). Another major problem was the lack of access to the relevant data due to issues of copyright and protectionisms (Lyytinen et al., 2016). Some companies and jurisdictions were reluctant to engage in anything that could be deemed to be a sharing economy (Lewis, 1996). Even where the findings were duly presented to the decision-makers, there was no guarantee that they would use them (Miller, 2014). Indeed, some of the decision-makers actively avoided incorporated data science in their calculus (Lewis, 1996). Reports were prepared at considerable expense and then not used at all (Mieczakowski et al., 2011). Not only does this represent a waste of resources, it also means that sub-optimal decisions are being made without the benefit of understanding the data science that would rationalize them (Miller, 2014).

Data analysts may have problems trying to explain the complex issues in a data set to laypeople and this means that they will not be able to properly influence the decision-maker (Hilbert and Lopez, 2011). At other times, the analyst is prevented from getting the data that they need because of concerns about privacy and data protection (Bachman, 2013). This can make the analysts overly cautious in their approach and reduce the depth of the research that they are undertaking (Helmreich, 2000). Smaller organizations often fail to take on board data science because their budgets are so restricted that they see no significant value in commissioning yet another service that they can ill afford (Little, 2002). In order to become a truly data-driven organization, a company has to understand the bottlenecks that prevent proper analysis from multiple angles that move beyond the restrictions on the internal activities of the organization (Hamari et al., 2015). For example, they can look at big data as a partner industry that has to be integrated within their own operations (Kees et al., 2015).

It is also important to recognize the limitations of traditional packages such as Microsoft Excel when dealing with big and complex data (Hamari et al., 2015). The organization must commission suitable software and hardware in order to support this type of functionality over the long run (Jibril and Abdullah, 2013).

6.3. SOCIAL STATISTICS AND SOCIAL ENTERPRISE

One of the clear applications of data science is that of social statistics and social enterprise (Awang et al., 2013). This is a successful amalgamation of the social sciences of political economic and social development (Gibson and Brown, 2009).

However, the statistical analysis is an important tool for making decisions (Helmreich, 2000). This is particularly true for those social entrepreneurs who may have very limited statistical training before they embark on entrepreneurship (Jibril and Abdullah, 2013).

Besides, it is important to understand that social enterprise has somewhat different priorities from traditional enterprise and that might change the type of statistics that are collected during business decision making (Mieczakowski et al., 2011).

For example, a traditional enterprise might consider the number of customers that it brings in and the marginal profits that are associated with each customer acquisition (Lyytinen et al., 2016).

In contrast, the social enterprise may want to understand the social cost of its business as well as its social benefits (Lewis, 1996). This is a rather more complicated layer of analysis that requires some specialization in order to achieve (Noughabi and Arghami, 2011).

Figure 6.3 show the social media as one of the major sources of big data is important for enterprise in our contemporary world.

Figure 6.3. Social media and enterprise.

Source: Hootsuite Blog.

Social enterprise has become so important now that even businesses that did not start off with a social ethos are considering including it in their publicity so as to earn the support of social consumers who are looking for ethical sound producers to engage with (Cappellin and Wink, 2009). Information is a key resource for social enterprises which may not be able to compete with traditional companies who are very effective in terms of cutting down costs and increasing revenue (Hair, 2010). The fact that customers are highly engaged with social enterprise creates an impetus for the entrepreneur to identify better ways of engaging with their clients at this level (Bachman, 2013). The BI that is used for decision making in this sector is critical because often the social entrepreneur does not have many other sources of good quality advice (Hair, 2010). By using data science in order to identify the needs and patterns associated with their target consumers, these social enterprises can deliver a bespoke service that is very attractive to their target customers (Holmes, 2005). It also means that social enterprises do not waste their resources on product lines or processes that are not going to give them the kind of foothold they need in the industry or segment (Lyytinen et al., 2016). Using social statistics intelligently can allow social enterprises to target those segments of the economy where they are most likely to have a positive impact rather than doing a generic service delivery that may not meet their social objectives (Hamari et al., 2015).

The use of social statistics is not just for the real-time management of the social enterprise. It can also be used to attract funding because things like feasibility must be accompanied by convincing information about the market and the positioning strategy that the social enterprise seeks to adopt when penetrating that market (Ellison, 2004). The availability of open science can be an advantage to social enterprises because they can tap into management information that is available for a low price or even free (Dutse, 2013). If the enterprise is involved in certain social projects, it can also produce information that can be shared with other potential partners (Little, 2002). Some of the larger corporations are unable to engage in social enterprise due to pragmatic reasons (Menke et al., 2007). However, these companies may be willing to partner with social enterprises (Lyytinen et al., 2016). The connection happens when social data came in to a particular social enterprise and its impact on the community as a whole (Dutse, 2013). That was the case when supermarket chains in the developed world started engaging ethical buying so as to ensure that farmers in the developing world were able to enjoy a fair share of their contribution to a particular production process (Howells and Wood, 1993). That impact was highlighted using social statistics at the local, national, and international levels (Mieczakowski et al., 2011).

There is a social sphere out there that is not yet fully explored (Jansen et al., 2008). The use of social enterprise and social statistics is meant to reach those spheres by using modern technology (Kim and Jeon, 2013). As we speak of co-production and co-consumption there is an attraction for consumers when they have information about the impact that social enterprises are making on the community as a whole (Holmes, 2005). As a planet, we are becoming increasingly aware of our social responsibility and this is particularly apparent in the fact that well over half of all US social enterprises were created in 2006 or later (Bachman, 2013). It is a trend that is growing in other countries. For example, more than 89% of social enterprises in India are less than 10 years old, with more than 57% of social enterprises in Canada founded in the past six years (Chiu et al., 2016). This is a young segment that needs all the publicity that it can get (Holmes, 2005). The use of social enterprise and social statistics can be one of the ways in which social issues can be brought to the fore as well as highlighting the efforts of those that are trying to do something about the status quo (Mieczakowski et al., 2011). Existing literature has highlighted the relationships between social enterprise and the knowledge economy (Hair, 2010). Many young people are drawn towards the knowledge economy and its approaches (Lewis, 1996). Social enterprise is very much part of this youth outreaches (Lewis, 1996). Data science can be the glue that links that two sides together (McFarlane, 2010). Besides, the interaction can also generate new data and knowledge which can help other sectors when they are making decisions about their own future (Mosher, 2013). In summary social enterprise and social statistics is part of the future knowledge economy (Mosher, 2013).

6.4. COMPUTATIONAL COMPETENCE FOR BUSINESS LEADERS

The development of computational competence for business leaders is a key strategic concern for any organization, regardless of whether it is in the public or private sector (Boase, 2008). Nobody can get away with not knowing anything about computers because this is the pervasive and predominant modality for business communication (Hamari et al., 2015). Computing has become a way of life and seeped into various arenas of our existence (Hair, 2010). Businesses use computing to collaborate and communicate. Individual customers use computing in much the same way, a conduit for all their transactions with others (Hilbert and Lopez, 2011). Virtually all academic disciplines have some level of computing (Ifinedo, 2016). The last decade has also seen the rise of disciplines generically described as "computational X," where "X" stands for anyone of a large range of fields from physics to journalism (Lewis, 1996). Computation is very much part of the curriculum, but it is not entirely clear whether the graduates of that education are fully functional in terms of optimizing their use of computing (Lyytinen et al., 2016). Figure 6.4 highlights some of the features of computational competence.

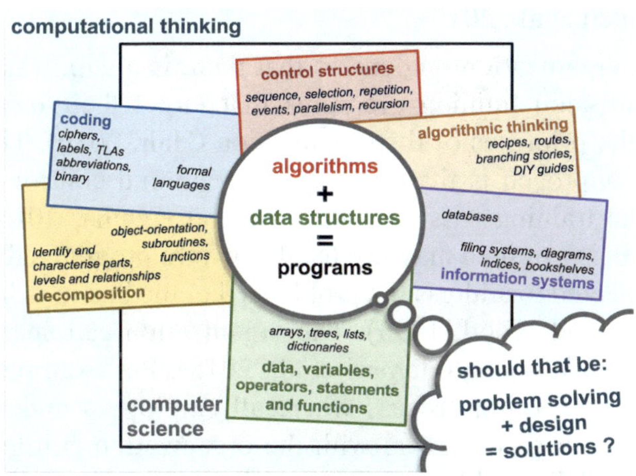

Figure 6.4. Elements of computational competence.

Source: Richard Millwood.

Business leaders may think that they no longer need computational pedagogy since they are experienced in their respective fields (Bachman, 2013). However, the age of big data has also demonstrated that the decision-making role of managers in any organization necessarily calls for some level of computational competence (Helmreich, 2000). Indeed, the requirements are not only for basic competence, but also the ability to be imaginative in the use of computational facilities so that they can better their decision-making (Kirchweger et al., 2015). There is always talking of the "4C's of 21st century" in existing literature (Carr, 2010; Hair, 2010; Jibril and Abdullah, 2013). These skills include critical thinking, creativity, collaboration, and communication (Awang et al., 2013; Hamari et al., 2015; Kees et al., 2015). The 4Cs have been widely recognized as essential ingredients of school curricula (Bansal, 2013; Hamari et al., 2015; Min et al., 2008). This shift has prompted an uptake in pedagogies and frameworks such as project-based learning (Gilks, 2016), inquiry learning (Howells and Wood, 1993), and deeper learning across all levels of formal education that emphasize higher-order thinking over rote learning (Bansal, 2013; Gilks, 2016; Noughabi and Arghami, 2011). Students are being encouraged to solve problems computationally (Evans, 2009). This means that they think algorithmically and logically about problems so that they can use computational tools in the problem-solving process (Howells and Wood, 1993). These tools may be used to create certain artifacts such as data visualization and predictive models (Lyytinen et al., 2016).

Where an organization recognizes that there is a gap, it is possible for them to commission training programs that target their own employees according to the priorities of the organization (Hair, 2010). The advantage of taking this approach is that there is direction and control over what is included in the training sessions (Malathy and Kantha, 2013). Similarly, those that are being trained may use local software programs that have been specifically selected to address the problems that the business has or its own goals (Howells and Wood, 1993). The organization can specify the pace and modalities of learning (Hamari et al., 2015). For example, they could create a lifelong learning process where all employees undergo refresher training throughout their careers with the organization (Little, 2002). The teaching can be delivered by a trained professional or it can be carried out by someone that is experienced through prior training (Gilks, 2016). The learning is complemented by frontline practice which is invaluable when dealing with complex concepts that do not make much sense until they are handled practically (Mieczakowski et al., 2011). Typically, computational competence for business leaders is not always tested during the recruitment process (Gibson and Brown, 2009). The assumption that these skills can be acquired on the job and in reality that is what happens for most organizations (Howells and Wood, 1993). Ironically, it is the low-level employees that are often required to demonstrate that they are computer literature (Jansen et al., 2008). It is important to note that computer literacy is not the same as computational competence (Mieczakowski et al., 2011). The former is at the more basic operations level while the other suggests some knowledge of analytical frameworks (Malathy and Kantha, 2013).

The more senior managers may feel that they do not need to acquire computational competences since they do not expect to be producing management information (Hair, 2010). However, that is the wrong attitude when trying to become a data-driven organization (Mieczakowski et al., 2011). In such organizations, every member of the organization is part of a unit relies on information and also produces information in return (Hair, 2010). The manager may not directly produce reports, but they have a responsibility for others that produce reports (Hilbert and Lopez, 2011). They are also supposed to make decisions that are based on the information that is gleaned from those reports (Little, 2002). Indeed, it may become a priority to ensure that all decisions have some element of data to back them up rather than relying on the discretion of the leaders alone (Little, 2002). The attitudes that the various workers have at all levels will reflect the type of organization that you are developing (McFarlane, 2010). The competencies should be shared among the various employees so that they are fully embedded with the organization (Holmes, 2005). When the competencies are integrated into the organization, it makes it easier to engage with potential partners who are also data-driven (Hamari et al., 2015). The sharing economy calls for cooperation among the different actors in an industry since not all of them are in direct competition (Ifinedo, 2016). Besides, the market might be big enough to accommodate all the people that are operating in the industry (Holmes, 2005). From a data-driven company, it becomes a data-driven industry (Mieczakowski et al., 2011).

6.5. THE LANGUAGE OF DATA SCIENCE

Just like any other established field of study, data science has its own language (Bachman, 2013). This is the means of communication that ensures all users of data fully understand what is being communicated (Dutse, 2013). They also have the means to respond with their own messages

(Ifinedo, 2016). The language is constructed from the tools, technologies, and methodologies that are associated with data science (Lewis, 1996). Like other languages, it also serves the role of identity and is part of the norms (Holmes, 2005). For example, when people talk about financial ratios and regression analysis; it is only those that have been exposed to data science before who will understand what is being referred (Lewis, 1996). Among the data analysts, this language because of a professional identity that gives them a sense of ownership of the output from their profession (Sinclaire and Vogus, 2011). The practice of data science requires the use of analytics tools, technologies, and programming languages to help data professionals extract insights and value from data (Hair, 2010). A recent survey of nearly 24,000 data professionals by Kaggle revealed that Python, SQL, and R are the most popular programming languages (Kees et al., 2015). The most popular, by far, was Python (83% used). Additionally, 3 out of 4 data professionals recommended that aspiring data scientists learn Python first (Kim and Jeon, 2013). These languages are not just about communicating among professionals, but also executing the various tasks that are associated with data science (Howells and Wood, 1993). Figure 6.5 highlights some of the benefits of learning R, one of the data science languages.

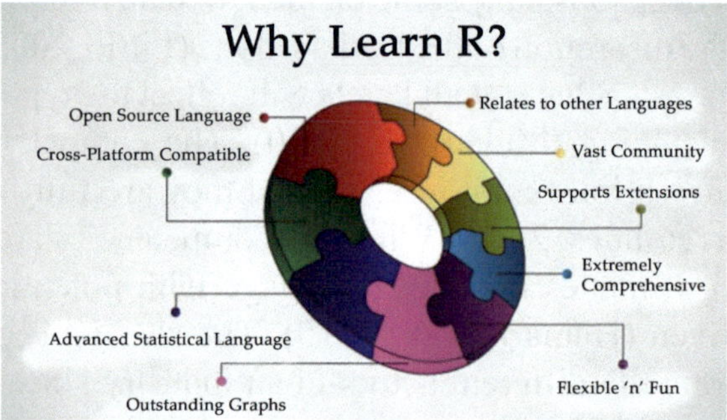

Figure 6.5. Language and data science.

Source: Data Bots.

Kaggle conducted a worldwide survey in October 2018 of 23,859 data professionals (2018 ML and Data Science Survey). Their survey included a variety of questions about data science, ML, education, and more (Helmreich, 2000). Kaggle released the raw survey data and many of their members have analyzed the data (Hair, 2010). The results indicated that data science languages have become a major determinant of whether or not organizations become data-driven (Howells and Wood, 1993). The professionals that seek to work within the data science industry are advised to learn one or more of the popular languages so that they are competent to handle the various tasks that may be assigned to them (Malathy and Kantha, 2013). When participants were asked about the most popular languages that were used in data science on a regular basis, Python was the top programming language were up to 83% of all the respondents indicated that this was a language that they used on a regular basis. A very distant second language was SQL with a high rating by 44% of the participants. The third most used language was R which was given top billing by up to 36% respondents. When designing a data management training program, these languages should be part of the consideration in terms of their ability to raise the prospects of the professionals that work in the data science industry (Jibril and Abdullah, 2013). It is worth noting that these data science languages are constantly evolving in order to account for the new demands that are made on them by the public (Chesbrough, 2005). Each organization has to do its own independent research in order to understand the specific languages that seem to be popular so that it can include them in its training programs (Jansen et al., 2008).

Learning the languages of data science is not a purely academic endeavor that has no practical implications (Jansen et al., 2008). A computer language is never truly mastered until it is used regularly (Howells and Wood, 1993). Indeed, those that do programming report that they do a lot of learning on the job rather than in the classroom that tends to go through a summary of the basic features of the language (Holmes, 2005). It is on that basis that every effort should be made to ensure that all members of the team are able to make use of the data science language in their work (Hilbert and Lopez, 2011). They may not be actually programming the activities, but they will be referencing the various functions on the system (Hilbert and Lopez, 2011). Understanding why the system does certain things is all part of the learning and will allow them to have a deeper understanding of how the system really works rather than merely relying on surface configuration or their meanings which do not explain why the data science is being produced in a particular way (Lyytinen et al., 2016). The evolution of data science language will also be linked to the evolution of the discipline itself since there is an interest in creating relevant support mechanisms for the corporate and non-corporate world in as far as decision making is involved (Malathy and Kantha, 2013). Existing literature has not yet critiqued the extent to which the language of data science is facilitating the embedded of the field in the business world

(Gilks, 2016). Anecdotal evidence suggests that although the vast majority of decision-makers are interested in what big data has to say, they are often put off by the language in which the reports are written (Malathy and Kantha, 2013). The language tends to veer between statistical and technological concepts which are beyond a layperson that is not steeped in the traditions of data science (Little, 2002).

Despite its influences on the discipline, it is tempting to suggest that data science is nothing more than a collection of statistical inferences and technological concepts which are then used to predict the future. In fact, data science is moving away from these narrow and linear definitions (Engelberger, 1982). Instead, data science is fully incorporated into the operations and corporate image of any organization, not least because data-driven businesses have a set of values that are espoused by their focus on the story that data is telling to them (Jansen et al., 2008). When designing curricula for data science, it is possible to include specific languages whose competency is an essential part of the professional requirements (Holmes, 2005). However, the practitioner should also be encouraged to be cognizant of the wider implications of data science in terms of supporting business development and organizational efficiency (Lyytinen et al., 2016). Typically, the training and development department will be actively involved in any training in the data science languages. In doing this work, the department should ensure that they set the curricula within the context of a complex organization that is working together in order to achieve specific goals (Holmes, 2005). The focus is not on the language of data science per se (Mieczakowski et al., 2011). Instead, it focuses on how that language can be used effectively in order to achieve the business goals (Hamari et al., 2015).

CHAPTER 6: SUMMARY

This chapter sought to suggest some areas that might be included in the data science curricula in order to help establish the discipline much more widely. The first section of this chapter showed that advanced probability and statistical techniques are an essential aspect of data management which is a core requirement for higher education in professional data science courses. The second section highlighted the importance of training staff members on the most up to date software packages which in turn support data science. The third section showed that social enterprise is a case in point of how data science in the form of social statistics can transform the prospects of a business entity even if it does not have a lot of opportunities to begin with. The fourth section suggested that all leaders and managers should endeavor to educate themselves about data science even if they are not directly involved in its production because this will help them make better decisions in their roles. The fifth section showed that there are many language types and categories that are emerged in order to describe various aspects of data science. It is recommended that these languages are included in any curriculum. The penultimate chapter in this book will consider the ethical issues that arise in data science.

CHAPTER 7

ETHICAL CONSIDERATIONS IN DATA SCIENCE

CONTENTS

This chapter highlights some of the ethical considerations and dilemmas that surround data science. The first section will highlight the requirements of data protection and privacy. The second section will discuss the nature of informed consent and primary usage. The third section will highlight the protective mechanisms required for storing and securing data. The fourth section will discuss the importance and possibilities of data quality controls. The fifth section highlights the need to protect business secrets as well as resisting unwarranted political interference in data science.

7.1. DATA PROTECTION AND PRIVACY

There is a lot of concern about how data science will impact on data protection and privacy issues (Awang et al., 2013). As more and more information is being curated about citizens, some are wondering the extent to which these private entities that are collecting the data are thinking about the safety and security of the subjects of that data collection (Hamari et al., 2015). Sometimes, it is not even the organization itself that has compromised the security of the citizenry (Little, 2002).

A company may be subject to a cyber-attack in which criminals access vital information which they can use to commit further crime (Mosher, 2013). The law as it stands generally allows for litigation against the company that has collected the data (Mosher, 2013).

That is why data protection and privacy issues in data science should be of concern to any organization (Gibson and Brown, 2009). Machine learning (ML) can sometimes present issues of privacy because it allows computers to effectively do the work that might have been done by a human being (Berker et al., 2006). The human being will have a sense of judgment and may even be concerned about personal liability if they end up breaking the rules of the game (Evans, 2009). A machine has no such ethical dilemmas and is designed to efficiently seek for information about its target, no matter how personal that information is (Holmes, 2005).

It is only when the information is passed on to a human analyst that they can take a discretionary to include or exclude data that seems to be an invasion of privacy (Ifinedo, 2016). The other issue is that technology has provided industry with the tools to blatantly and sometimes surreptitiously break privacy (Sakuramoto, 2005). In the UK, the phone-hacking scandal exposed the risks that celebrities faced when their movements were clandestinely monitored by tabloid newspapers in order to diver outlandish narratives in the press (Berker et al., 2006).

The corporate governance framework for any organization ought to address the data protection and privacy issues in data science (Gilks, 2016). This is because many of the perceived failings in this area are directly linked to the collection, manipulation, and presentation of big data (Holmes, 2005). Sometimes the people how privacy is being invaded do not even realize that their information is being collected and shared (Ifinedo, 2016). The people that collect the data prey on the fact that users typically get irritated when their online surfing is interruption by contractual documents and detailed consent forms (Helmreich, 2000). Hence, the user instinctively clicks to accept all the terms and conditions without reading them because all they want to do is get to their destination and see the content on the page that they have searched for (Menke et al., 2007). Taking that as consent, the person or entity that is collecting data will then monitor virtually all the movements that are taking place and record them in databases which can be accessed for commercial purposes (Menke et al., 2007). Not all organizations take such a cavalier attitude towards issues of privacy. Figure 7.1 is an example of global privacy values that can prevent the exploitation of consumers in this way.

OUR GLOBAL PRIVACY VALUES

RESPECT	TRUST	PREVENT HARM	COMPLY
We recognize that privacy concerns often relate to the essence of who we are, how we view the world and how we define ourselves, so we strive to respect the perspectives and interests of individuals and communities and to be fair and transparent in how we use and share information about them.	We know that trust is vital to our success, so we strive to build and preserve the trust of our customers, employees, patients and other stakeholders in how we respect privacy and protect information about people.	We understand that misuse of information about people can create both tangible and intangible harm for individuals, so we seek to prevent physical, financial, reputational and other types of privacy harm to individuals.	We have learned that laws and regulations cannot always keep pace with the rapid change in technologies, data flows, and associated shifts in privacy risks and expectations, so we strive to comply with both the spirit and the letter of privacy and data protection laws and regulations in a manner that drives consistency and operating efficiency for our global business operations.

Figure 7.1. Example of a data protection policy framework.

Source: MSD Responsibility.

The ubiquitous use of the mobile app to collect data about people and events raises ethical concerns because there are very limited controls on what is collected and how data is eventually used (Engelberger, 1982). Personal information must remain exactly that and should only be used for commercial purposes with the explicit and informed consent by the person that is being monitored (Helmreich, 2000). During the 2016 US presidential election, there was concern that a company called Cambridge Analytica (CA) that specialized in consulting for large election campaigns was actually profiling voters using information that was gleaned from an unrelated psychology research project.

The data that the researcher obtained from the psychology experiment was actually sold for $200 million, none of which ever entered the pockets of any of the subjects of this data collection. These are the types of data protection and privacy issues in data science which cause the public to treat the entire scheme with suspicion (Ifinedo, 2016).

7.2. INFORMED CONSENT AND PRIMARY USAGE

We have already touched on the issue of informed consent (Davis et al., 2014). The data science industry has not always been ethical or considerate in this matter (Holmes, 2005).

The person whose data is about to be collected is manipulated into giving consent without actually understanding the implications that their consent has for their privacy and safety (Lewis, 1996). Academic researchers are typically required to address the issues of informed consent and their studies will not be published in respectable journals if they have not met this requirement (Ruben and Lievrouw, n.d.).

However, big data is being analyzed by non-academicians who have far less stringent controls on consent (Bansal, 2013; Schute, 2013). That is why some private companies do away with consent altogether and merely collect the data (Hilbert and Lopez, 2011;

Noughabi and Arghami, 2011). There are growing discontinuities between the research practices of data science and established tools of research ethics regulation (Menke et al., 2007; van Deursen et al., 2014). Some of the core commitments of existing research ethics regulations, such as the distinction between research and practice, cannot be cleanly exported from biomedical research to data science research (Awang et al., 2013; Berker et al., 2006; Hilbert and Lopez, 2011;

Sinclaire and Vogus, 2011). Figure 7.2 shows how the UK government has incorporated marketing practices into its consumer protection mechanism. This sets the stage on which consumers are not subjects of exploitation but active participants in a business relationship.

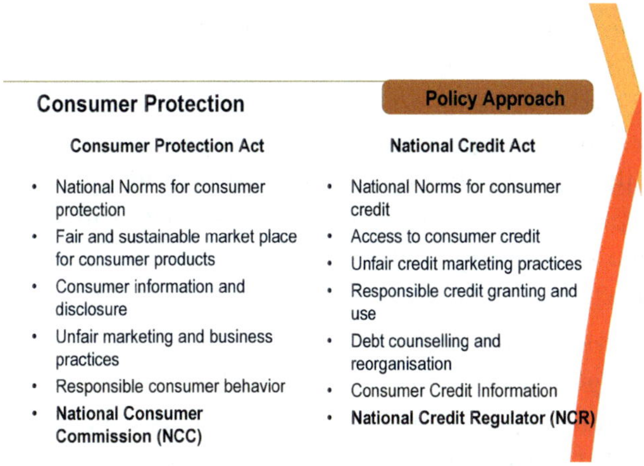

Figure 7.2. Regulatory framework for consumer protection in the UK.

Source: The Department for Trade and Industry.

One of the contentious issues that have emerged is the rejection of official regulation by the state (Hilbert and Lopez, 2011). Whereas academicians have always had to account for their ethical behavior in research, data science practitioners are not yet placed under the same stringent regime (Bachman, 2013). That means that they can occasionally bend the rules (Ifinedo, 2016). Regulations such as the common rule which were accepted as the best practice module in the USA are under review (Menke et al., 2007). Big data and data analysis has created new ethical dilemmas that would have been virtually absent in the pre-internet era (Min et al., 2008). For example, we now know that it is virtually impossible to completely erase something that is put up on the internet (Miller, 2014). The regulatory framework for ethics has also been somewhat harmed by the disparate operations of the various regulatory body (Hair, 2010). Each discipline will set out its own standards and some of them do not coordinate with others in order to have uniform standards (Jibril and Abdullah, 2013). However, the issue of informed consent has always been an important lynchpin of academic research (Miller, 2014). The challenge is to make it similarly relevant to non-academic research which is being used by different decision-makers in the private sector in the era of big data (Noughabi and Arghami, 2011). Another issue that is of concern is the fact that big data is collected for different uses from the ones that it actually performs (Hilbert and Lopez, 2011). Sometimes the subjects of that data collection are not even aware that they are part of a data experiment or that their information can be used in perpetuity by some as-yet-unknown entity (Gilks, 2016). Others may not be aware that their data is being commercialized without their consent or that the consent that they are providing has far-reaching consequences beyond the click that they use to signify assent (Jansen et al., 2008). It is not even clear that the public wants to be informed about the implications of data collection since they are typically captured in the middle of a transaction which is prioritized over consent issues (Mosher, 2013).

During the polarizing debates about privacy and security online, an idea merged of global consent the moment that someone signs up to any information technology network (Hair, 2010). This is based on the premise that it is common knowledge that the internet is not very secure and that information that is captured while surfing can be used in different contexts (Helmreich, 2000). The implication is that the internet is such a bad place for privacy and security that anybody that joins it is in effect falling under the "buyers beware" category (Gibson and Brown, 2009). The problem with such a supposition is that it takes away the right of consent from the user and assumes that they are willing to have their data commercially exploited by virtue of the fact of using a facility that is ideally open to the public (Hamari et al., 2015). Besides, the logical conclusion to that argument is that people who value their privacy and rights should stay away from the internet (Helmreich, 2000). Such an outcome will mean that potential consumers are being driven away from the online market, simply because the actors on that platform are unable to or unwilling to engage in commonsense self-regulation (Ifinedo, 2016). In response to the enduring problems of informed consent and primary usage, some disciplines have proposes some flexible regulatory frameworks that can encompass both biomedical and non-biomedical research (Chiu et al., 2016). There is a focus on public data sets which mainly do not have informed consent prior to collecting the data, but are nonetheless widely used in the industry as reference points (Kirchweger et al., 2015).

Organizations have to make a decision as whether they will make use of data sets whose collection has not obtained completely informed consent (Hair, 2010). If they uphold the stricted ethical considerations, they may be left in a position as the only firm in the industry which is not accessing the big data that is publicly available (Holmes, 2005). Others may opt to contact participants directly and obtain informed consent (McFarlane, 2010). However, the reality of that consent is not clear-cut since most of the time online readers are not interested in following long text of legal precedents and contractual obligations when they are in the middle of seeking a particular page or product (Min et al., 2009). In terms of primary use, individual organization have very little control over the big data that is publicly available and used by different firms in order to strengthen their commercial position (Lyytinen et al., 2016).

7.3. DATA STORAGE AND SECURITY

Data storage and its security are some of the practical concerns for any organization that is involved in using the output from the data science industry (Bachman, 2013). Given the increasing security risks online, some companies have created the position of a data security scientist (Gilks, 2016). This is a professional that is takes with identifying risks and then coming up with mitigation plans that can control or eliminate those risks (Kees et al., 2015). The position will sit within the information technology team but may be provided with a wider remit depending on how the organization wishes to structure its teams (Helmreich, 2000). The person that is handling this position will have a lot of direct and indirect power (Ifinedo, 2016). Therefore, it pays to ensure that recruitment, training, and development are carefully calibrated to get the best personnel in this role (Ifinedo, 2016). Meanwhile, senior managers have to exercise oversight role by regularly checking to ensure that the issues of data storage and security have been adequately addressed by the present arrangements (Mieczakowski et al., 2011). All this is part of data management (see Figure 7.3).

Figure 7.3. The importance of data management.

Source: Blue Pencil.

The data security officers are a new breed of data scientists with a specific remit that is supposed to protect the integrity of the data management process (Awang et al., 2013). The more data an organization deals with, the greater the likelihood of need such posts such as these (Holmes, 2005). In order to protect the data, the scientist must develop a broad understanding of the intricacies that are involved when handling such data (Helmreich, 2000). It may require them to liaise with other members of the team so that they can map their own experiences of mapping the data (Ifinedo, 2016). The assumption is that the resultant improvements will enhance decision-making and it makes worthwhile to recruit a data security officer (Little, 2002). Some jurisdictions even require such a post to act as a contact for the regulator when they wish to ascertain the kinds of steps that have been taken in order to protect the consumer when collecting large databases (Kirchweger et al., 2015). The post should be fully integrated into the daily activities of the organization based on the notion that work that is accomplished is integral to the operations of the business and is not merely a technical offshoot of activities (Helmreich, 2000).

The data security officer may take on the role of accelerating access as well as building alliances with other organizations that are in the same data loop (Gibson and Brown, 2009). Of course, this calls for some harmonization of systems and objectives so that the partners are not working against each other's interests (Gibson and Brown, 2009).

The recruitment of such an officer is recognition of the gradual convergence of information technology, data management and security issues in the age of online transactions and egalitarian approaches to data usage (Jansen et al., 2008). Social media, in particular, offers great opportunities for interacting with members of the public and potential business partners (Hilbert and Lopez, 2011). However, social media has also been the source of many data breaches and there is always a requirement to ensure that security issues are at the forefront of all the planning that goes into social media campaigns (Helmreich, 2000). Including data scientists in security decisions has the potential to bring long-overdue disruption to information technology security departments (Gilks, 2016). This shows that offers must work in a collaborative way and learn from the techniques and approaches which are utilized in other departments (Jibril and Abdullah, 2013). In much the same way as data scientists improve business decisions; data security offers are able to play a similar role and even facilitate the means through members of the team access data (Mosher, 2013).

Another radical approach that is used by some organizations is to have a centralized data security team that provides services and oversight to the rest of the organization (Hilbert and Lopez, 2011). The downside for that is that the expense of setting up and running a team of that magnitude might be too much for the types of small businesses that are looking to join the data-driven economy (Menke et al., 2007). In any case, all employees should be aware of security issues as they pertain to the way in which the organization utilizes data science (Hair, 2010). In this way, it is possible to take the necessary precautions on an individual basis (Miller, 2014). This will help to embed the notion of personal responsibility for the way in which data is used (Holmes, 2005). When data protection legislation as introduced in some developed countries, the approach that included personal responsibility ensured that, the entire organization was broadly committed to the ethos and ethics that underpinned the act (Lewis, 1996). Although the penalties for non-compliance were mainly directed to the corporation as a whole, every foot soldier would be involved in ensuring that that eventuality never happened (Little, 2002). The same approach could be used for issues of data security (Wallace, 2004). In any case, the failure to properly use big data could have consequences for the viability of the organization which in turn affects the job prospects of its workers (Howells and Wood, 1993). That means that in effect everybody has a vested interest in ensuring that the system works (Evans, 2009). Some of the data breaches occur because the people that are on the frontline are not fully informed about their responsibilities and the possible impact of their failure to undertake those responsibilities (Holmes, 2005). It is a mistake to make data security the exclusive concern of a few select employees when in reality it is something that can have a profound impact on the entire organization (Ruben and Lievrouw, n.d.). This is one of those situations in which the notion of co-production and co-consumption of big data for decision making can be applied internally within the organization in order to make it more effective (Jibril and Abdullah, 2013).

7.4. DATA QUALITY CONTROLS

The decisions that are made as a consequence of data science rely largely on the quality of the data that was used (Dutse, 2013). That is why data quality controls have been incorporated into the wider framework of data management (Jansen et al., 2008). The technical term that is used in existing literature is data quality management, or DQM (Helmreich, 2000). It consists of set practices that are designed for the specific purposes of ensuring that the information which is used to make decisions is of the highest quality possible (Kees et al., 2015). This approach may help to alleviate some of the concerns about the methodologies of data science when it is undertaken by people who are not trained academics (Kirchweger et al., 2015). The company that engages in data quality management is protecting itself from some deficiencies in the data that they are receiving (Noughabi and Arghami, 2011). Indeed, if this is an open resource that is shared with competitors, the fact the data is cleaned beforehand will provide the company with a competitive advantage (Min et al., 2009). Figure 7.4 demonstrates a six-step process for data quality management.

Figure 7.4. Data quality management framework.

Source: Digital Transformation Pro.

It is imperative to think of data quality management as a comprehensive process rather than a single event (Davis et al., 2014). It does not stop at the acquisition stage (Hilbert and Lopez, 2011). Instead, all the people that engage with the data will try to improve it so that it is more precise for purposes of making data-driven decisions (Min et al., 2008). These processes can be turned into policy documents that guide all the people that operate within the organization so that they are aware of the procedures to follow if there is any concern about the quality of the data that they have encountered during the course of their job role (McFarlane, 2010). Indeed, data quality management can extend to the distribution of data so that it is sent to the right places at the right time (Little, 2002). Quite often decision-makers ignore irrelevant data and that can be a serious problem if it habitually happens (Helmreich, 2000). The people that are responsible for the quality controls should also monitor the distribution channel in order to ensure that it is still fit for purpose, regardless of the kind of expansion that the organization has gone through (Kirchweger et al., 2015). When data quality management is done well, it can improve the insights that are gleaned from information and therefore make for good decision-making.

Each company tends to develop its own approach and methodologies for data quality management. Some organizations will hire officers or other staff the clean the data before it is analyzed (Dutse, 2013). Alternatively, cleaning the data can be part and parcel of the analysis so that each decision-maker can spot the mistakes and inconsistencies long before they start the analysis (Helmreich, 2000). More importantly, managers should be inquisitive enough to query the information that is provided to them just in case it is not based on sound data science (Mosher, 2013). It takes considerable skill and some experience to be able to take on an analyst when they are presenting data that does not add up (Min et al., 2008). For some managers, it is simply easier to accept whatever is being given to them and take it for granted that the person advising them is an expert whose professional ethics would not allow them to present bad data (Lyytinen et al., 2016). Although this is the case in the vast majority of cases, there are instances where the analyst may miss important aspects of the analysis (Howells and Wood, 1993). The resultant decisions can have significant effects on the organization (Gibson and Brown, 2009). It is much easier to prevent those mistakes happening than trying to deal with the consequences of those mistakes at a later date (Lewis, 1996). Because data quality management is a process, it can be built within operations so that it happens as a matter of course (Hamari et al., 2015). Each decision-maker continuously checks the information that they are working with (Jibril and Abdullah, 2013). They can engage in triangulation if they feel that there is over-reliance on one set of data (Lyytinen et al., 2016). Even as they query the problems within the data, the analyst is learning more about the data and will, therefore, improve their overall competence (Sinclaire and Vogus, 2011).

There are many arenas in which data quality management remains relevant (Hair, 2010). These include customer relationship management (CRM) where the people that are purchasing the output from the company are looking to get the best service possible (Kirchweger et al., 2015). One of the competitive advantages for an organization will be if it is so well informed that it is able to provide bespoke services to its customers as compared to other competitors who may adopt a generic approach (Kees et al., 2015). The implementation of a data quality management framework can also have some positive impact on the supply chain management within the industry (Mieczakowski et al., 2011). Suppliers and buyers feel more confident about a partner that is always ready with the right goods and services when called upon to do so (Awang et al., 2013). Similarly, a well-run supply chain will ensure that the production line is not disrupted due to miscommunication (Boase, 2008). In fact, data quality management is part of enterprise resource planning (ERP) because ultimately the data also references the resources that are available or needed for that particular organization in order to run its operations (Helmreich, 2000). It is not surprising that the benefits of effective data quality management can have a ripple impact on an organization's performance (Holmes, 2005). With quality data at their disposal, organizations can form data warehouses for the purposes of examining trends and establishing future-facing strategies (Menke et al., 2007). Industry-wide, the positive ROI on quality data is well understood (Ifinedo, 2016). According to recent big data surveys by Accenture: 92% of executives using big data to manage are satisfied with the results, and 89% rate data as "very" or "extremely" important, as it will "revolutionize operations the same way the internet did." Overall, data quality management must be an inclusive process that uses people, structures, systems, and software in order to achieve the strategic goals of the organization (Bansal, 2013; Kirchweger et al., 2015; Miller, 2014; Zhang and Chen, 2015).

7.5. BUSINESS SECRETS AND POLITICAL INTERFERENCE

In democratic countries, businesses are free to engage in commercial enterprise within the rules and regulations of the jurisdiction (Engelberger, 1982). Some of these regulations may reference health and safety while others talk about taxation and corporate social responsibility (Howells and Wood, 1993). However, there are parts of the world where business is wedded to the politics of that particular community (Holmes, 2005). For example, the wealthiest entrepreneurs may be an exclusive club of donors to the ruling political party (Ifinedo, 2016). The government calls in favors from private enterprises which in turn expect special treatment (Little, 2002). The value of data science has meant that some governments are beginning to take an interest in what happens as organizations are processing this data (Gibson and Brown, 2009). For example, China has strict policies on opening up businesses and the government regularly monitors communication (Awang et al., 2013). There are search engines which are banned from China and that means that overall data that is available for decision-making may be somewhat restricted (Schute, 2013). For businesses, these are obstacles that they have to navigate on a regular basis if they want to still remain operational in a particular jurisdiction (Schute, 2013). That is not to say that government is not a legitimate stakeholder in business or that business has no means of responding to that interference (see Figure 7.5).

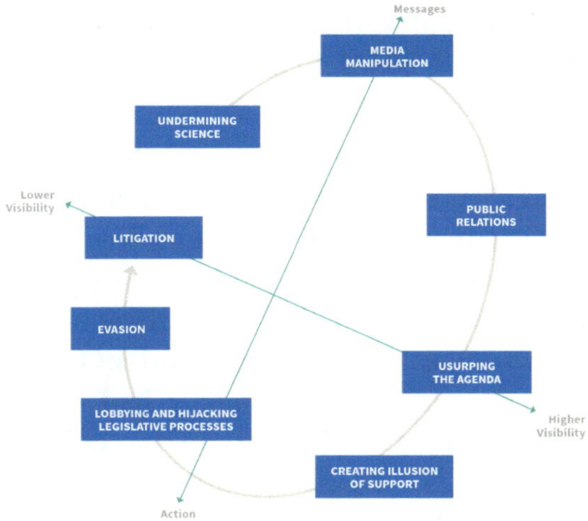

Figure 7.5. Government and international business.

Source: The Tobacco Map.

The fact that private companies are maintaining large databases of information about the citizenry is a matter that is of concern to government and one that might stimulate the government into creating legislative arrangements that are meant to keep controls over all these activities (Gilks, 2016). The problems arise if the government is working in ways that are harmful to private enterprise (Ifinedo, 2016). The companies themselves that rely on big data have to engage in certain preventative actions in order to avoid attracting government attention (Bansal, 2013). For example, they may ensure that they pay all their taxes on time and that they follow the provisions relating to consumer protection (Holmes, 2005). Of course, ethical dilemmas arise where the state makes it impossible to do business unless the company is a political donor to the dominant political party in the administrative (Holmes, 2005). This is a problem that confronts much organization that operates internationally (Ifinedo, 2016). At other times, the companies are punished by their home governments for aiding and abetting corrupt practices when they decide to engage in survival tactics that involve bribing the reigning regime (Malathy and Kantha, 2013).

There have been examples of corporations getting into trouble because of third party threats or even failing to follow all the regulations that are put in place by the government (Hilbert and Lopez, 2011).

Far more than 87m Facebook users' data may have been compromised as implications of big data combined with micro-targeting are only beginning to be understood (Chiu et al., 2016). In key exchanges in the UK and Ireland, the specter of more political interference from beyond borders using big data reappeared and could haunt Facebook for some time.

This is a brand that has been accused of accumulating large amounts of personal data from the citizenry but not doing nearly enough to protect that data from mischief by its own staff or third parties (Lyytinen et al., 2016). After what was a revealing day for Facebook, there are fears that CA may have had more quizzes in the wild. In the UK, former CA employee Brittany Kaiser revealed that the political consultancy had a suite of personality quizzes designed to extract personal data from the social network, of which Aleksandr Kogan's This Is Your Digital Life was just one example. Kaiser wrote in evidence to the House of Commons' digital culture, media, and sport select committee:

"The Kogan/GSR datasets and questionnaires were not the only Facebook-connected questionnaires and datasets which Cambridge Analytica used. I am aware, in a general sense, of a wide range of surveys which were done by CA or its partners, usually with a Facebook login—for example, the 'sex compass' quiz. I do not know the specifics of these surveys or how the data was acquired or processed. But I believe it is almost certain that the number of Facebook users whose data was compromised through routes similar to that used by Kogan is much greater than 87m, and that both Cambridge Analytica and other unconnected companies and campaigns were involved in these activities."

The complaints are not just about negligence, but also outright bias. President Donald Trump of the USA has consistently complained about the perceived bias against him and all political conservatives where their posts are classified as hate speech which needs to be deleted. Others fear that the owners of big data are harvesting vital information which may be used to influence political decision making (Hamari et al., 2015).

For example, media disinformation which is colloquially known as "fake news" has been implicated in electoral upheavals in both Europe and North America. Once politicians realized the power of big data, it was only a matter of time before they would start using it for purposes of swaying people to their side (Ifinedo, 2016). It is now impossible to decouple politics form big data (Gilks, 2016). The big challenge for those working in this industry is to ensure that the politicians do not compromise the professional standards that consumers expect when benefitting from a data-driven economy (Malathy and Kantha, 2013).

CHAPTER 7: SUMMARY

This chapter sought to highlight certain ethical considerations that are pertinent in the era of data science. The first section in the chapter showed that data protection and consumer privacy are competitive advantages for companies that do them properly, but also significant disadvantages for those companies that adopt a cavalier attitude towards them. The second section showed that it was very difficult to obtain iron-clad reassurances about informed consent and primary usage when dealing with data science since most of the information is collected in the middle of transaction and little effort is made to ensure that users understand the implications of their consent to have their personal information included in big data. The third section showed how data storage and security can help to improve decision making by virtue of ensuring that data quality is not compromised before it is processed into practical and useful information. The fourth section showed that data quality controls have to be an integral aspect of data-driven decision making since data is often corrupted throughout its life cycle and some of the problems are not discovered until later. The fifth section showed how the value of data science has inspired politicians to take an interest to it and sometimes this is to the detriment of the discipline. The upcoming final chapter in this book will explain the various ways in which data science has supported business decision making.

CHAPTER

8

HOW DATA SCIENCE SUPPORTS BUSINESS DECISION-MAKING

CONTENTS

This concluding chapter returns back to the core theme of the book by summarizing the different ways in which data science supports business decision-making in any given organization. The first section will explore how data science can open up the perspectives of decision-makers. The second section will consider how data science supports the evaluation of possibilities for decision making and feasibility studies. The third section will explain how good data science can justify decisions and provide a convincing rationale for stakeholders. The fourth section will explore the role of data science in fostering a culture of maintain records. The last section will demonstrate how the effective use of data science can improve the objectivity of decision making. The overall aim of this chapter is to leave readers with a take-home message that emphasizes the possibilities of modern data science at its best.

8.1. OPENING UP THE PERSPECTIVE OF THE DECISION MAKER

The starting point is the decision is routinely made based on the knowledge and experience of the person that is making the decision (Awang et al., 2013). This is actually the wrong approach in the age of data science (Hilbert and Lopez, 2011). There is so much information coming out, that it seems a waste if it is ignored in favor of what the decision-maker things they know (Evans, 2009). A case in point is how a recent advert for an off the road car raised alarm among female presenters because it was dominated by rugged men. It seems that the advertiser was thinking about women in the 1970s rather than in 2019. Such decisions are based on the failure to properly analyze and utilize the information that data science is providing to the decision-maker (Little, 2002). It is highly recommended and even essential that the decision-maker is always open to listen to the data science (Kim and Jeon, 2013). From that, they can start to weigh all the options before settling on the ones that seem to achieve their business goals (Holmes, 2005). If they refuse to acknowledge the contribution of data science, they are actually shortchanging themselves (Min et al., 2008).

Another mistake that decision-makers commit is to perceive data science from one perspective or one disciplinary framework (Bachman, 2013). Big data has an advantage in as far as it comes from a multiplicity of sources that should enrich the basis on which business decisions are made (Evans, 2009). The ability to access and interpret this data for optimum effect on the corporate goals is what is known as decision intelligence (Kees et al., 2015).

It is a phenomenon that is even present in the animal kingdom. For example, lions must glean very many aspects of their environment whilst on the hunt in order to foil the prey's effort to escape them. Academia has also started to take an interest in decision intelligence in order to understand why some executives are so much better than others when it comes to making critical decisions (Carlson, 1995). Decision intelligence is actually a multifaceted concept that is studied from a multidisciplinary perspective which includes social sciences, applied data science, management, and even economics (Ellison, 2004). This is in effect one of the vital sciences today which has a number of applications including contributing to the development of artificial intelligence that is uniquely geared towards the needs of the organization (Gilks, 2016). The issue that might call for specific attention is the recruitment of staff members that are competent enough to handle the more complex aspects of the data science (Gilks, 2016). Even where there is a skills gap, it can be covered using translation skills (Kobie, 2015). Some researchers actually suggest the data science can help decision-makers to set goals, objectives, and metrics that will be used to assess performance (Kobie, 2015). Data science today is mainly automated and therefore one of the tasks will be to ensure that there is harmonization of systems during the transitional phase (Kirchweger et al., 2015) (Figure 8.1).

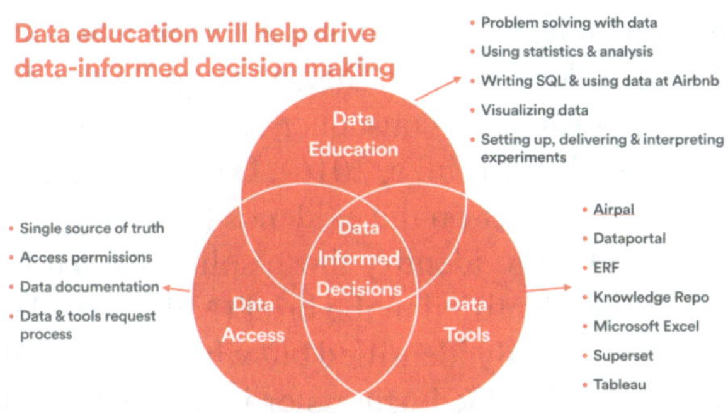

Figure 8.1. Using data science to educate decision makers.

Source: Decoded.

Decision intelligence is the discipline of turning information into better actions at any scale (Dutse, 2013). A number of firms may have access to the same information, but it is only the truly competitive ones that will know how and when to use it (Hamari et al., 2015). The data deluge means that there is plenty of information out there and that it is accessible to those that are willing to search for it (Helmreich, 2000). The companies that decide to pay for their analysis will get a somewhat better picture than those that rely entirely on open science outlets (Hilbert and Lopez, 2011). Regardless of whether or not the data is paid form, there is an impetus to engage in strategic thinking about which data is important and how it is important (Kim and Jeon, 2013). Producing or reading substantial reports that have little relevance to the decision-making scenario is counterproductive because it wastes resources (Kirchweger et al., 2015). The real data that needs to be worked on may even be hidden behind the various technical aspects that are picked into the report (Kirchweger et al., 2015). The strategic organization will ruthlessly discard that information that is not relevant or outdated in favor of current relevant information that can be used for forecasting (McFarlane, 2010). Overall, data science has played a critical role in opening up the perspectives of decision-makers in a variety of organizations (Ifinedo, 2016). It does not matter whether they are the most senior executive in the organization or a lowly support worker. Data science is still relevant to them and it can make them a more effective employee (Ruben and Lievrouw, n.d.).

8.2. PROPERLY EVALUATING FEASIBLE OPTIONS

In order to make good decisions, one must be able to evaluate the alternatives before selecting the ones that are most workable in the circumstances (Awang et al., 2013). Unfortunately, there are many decision-makers that are simply in a rush to get things done (Evans, 2009). They do not ruminate on the choices available to them and as a consequence make hasty decisions that are not supported by the evidence (Gilks, 2016). Data science can serve the role of focusing the decision-maker on the evidence (Jibril and Abdullah, 2013). They are then in a position to identify those solutions that best address the problems that have been identified in the brief (Miller, 2014). There may be many options that are not easily identified but which are actually quite useful for solving the problems that the business or individual encounters (Menke et al., 2007). The most successful modern businesses have embraced big data because it contains some of the options that were previously hidden from them (Holmes, 2005). Indeed, some of the data may be provided on an open science basis which does not cut into the bottom line (Jibril and Abdullah, 2013). This means that the company only needs to identify those information strands that are relevant to it and then use them accordingly (Noughabi and Arghami, 2011). Startups, in particular, can benefit from this arrangement because they tend not to have a large research and development budget (Jansen et al., 2008). Figure 8.2 demonstrates how data engineering and data processing can contribute positively to decision making in a firm.

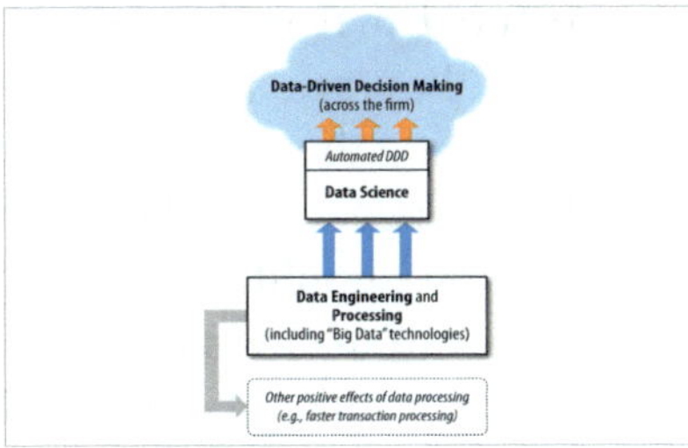

Data science in the context of various data-related processes in the organization.

Figure 8.2. Data engineering and processing in decision making.

Source: Sung-Soo Kim.

Expanding the options on which a final decision is made can bring about many benefits (Chiu et al., 2016). First of all, it allows the business to leverage its wealth or investment funds appropriately based on those projects that are most likely to yield a high return with manageable risk (Gilks, 2016). The fact that data science is digitized means that this information is at your fingertips (Hilbert and Lopez, 2011). Stockbrokers have embraced data science for this reason because it opens up their ability to creatively configure an investment portfolio for purposes of maximizing the income that an investor can get (Jansen et al., 2008). At the same time, this approach allows them to calculate and mitigate risks which can dent any profits that are made (Jibril and Abdullah, 2013). It is collectively known as business intelligence (BI), but so many businesses fail to take full advantage of the benefits that data science can bring to their decision making (Kim and Jeon, 2013). Businesses that consider all options are able to attain commercial growth in a way that is sustainable by leveraging their best resources to target the most lucrative opportunities that exist within their environment (Evans, 2009). This is not something that is only experienced when dealing with the external environment (Little, 2002). It can also be used to make decisions about the organization and re-organization of internal departments for the purposes of improving the bottom line and achieving organizational evolution (Mieczakowski et al., 2011). For example, there may be departments that are not really performing well and need to be put under certain measures in order to get the best out of them (Ulloth, 1992).

In order for all these benefits to be realized, organizations must implement the right reporting tools (Evans, 2009). They can monitor and evaluate them on a regular basis in order to identify where the data gaps are

(Helmreich, 2000). The personnel that make decisions as well as prepare reports must be trained so that they are able to optimize their analysis in ways that are conducive to achieving the stated business goals (Howells and Wood, 1993). Accuracy and other aspects of data quality are of the essence (Mosher, 2013). The organization should set aside a budget for ensuring that the data on which decisions are made is of the highest possible quality (Trottier, 2014). Existing literature talks about tangible insight as a much better approach to business decisions making than gut instinct (Engelberger, 1982). The tangible insight arises from knowing what is really going on and how it affects your business (Hamari et al., 2015). It is not a theoretical construct without practical value. Indeed, some of the more advanced data analyses allow you to engage in scenario mapping so that you can compare the expected outcomes of each type of decision that you make (Hilbert and Lopez, 2011). Businesses that have adopted tangible insight as the default decision-making model are more likely to succeed and sustain their success than those that are merely founded on gut instinct alone (Kobie, 2015).

Besides, it can be problematic identifying the person with the best gut instinct (Hilbert and Lopez, 2011). Typically, the owner will take on this role despite the fact that they may not have as much information about the business environment as other members of the team (Noughabi and Arghami, 2011). It is an arbitrary way of making decisions and in most cases might turn out to be detrimental to the prospects of the business (van Deursen et al., 2014). This is not about removing the entrepreneur from the decision-making process (Holmes, 2005). Rather, it is about streamlining and cleaning up the decision-making process so that it is rational and can withstand the test of time (Kees et al., 2015). Indeed, those businesses that have adopted a data-driven approach to decision making can build resilience through carefully studying the consequences of past and present decisions (Jibril and Abdullah, 2013).

8.3. JUSTIFICATION OF DECISIONS

Some business leaders have adopted a managerial style that is closer to an empire than a fully functioning business process (Hamari et al., 2015). They make all the decisions and are not accountable to anyone (Helmreich, 2000). This is particularly true where the chief executive officer of an organization is also its owner (Little, 2002). Small and medium-sized businesses fall into this category (Miller, 2014). Because they do not have to report to anyone about the decisions that they make, these leaders mistakenly believe that they need not justify their decisions (Ulloth, 1992). In fact, all business decisions must be justified by a rationale that is based on business data (Wallace, 2004). It is even better if these justifications are written down so that those who inherit the roles can understand what happened and the consequences of what happened (Bansal, 2013). Making justified decisions can bring about investment capital inflows since potential investors are attracted to those entities that are very clear about why they are taking particular courses of action (Holmes, 2005). This is different from businesses that typically look inward for capital financing and do not feel that they have to be accountable to their creditors and investors (Menke et al., 2007). Once again, it is a disease that afflicts the small and medium-sized businesses that tend to also have a high failure rate (Min et al., 2008). Figure 8.3 shows how the data science processes can take a company towards decision making.

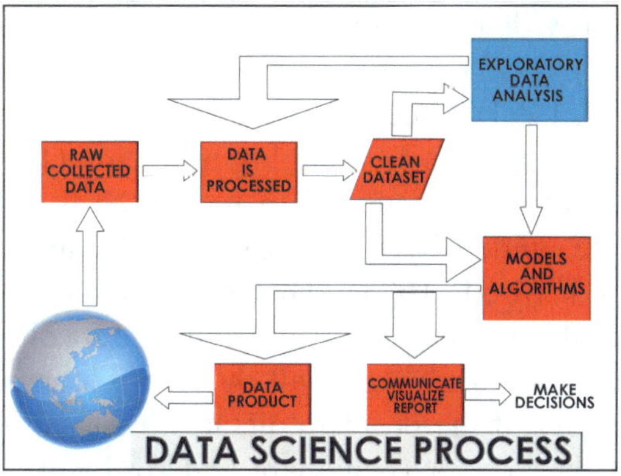

Figure 8.3. The data science process and decision making.

Source: DMAI.

One of the excuses that are sometimes provided by entrepreneurs when they do not want to justify their decisions is the fact that they have to make quick decisions which do not allow for reflection (Helmreich, 2000). However, that is a self-created crisis (McFarlane, 2010). There is plenty of big data out there and if the organization gets into the habit of regularly checking that data, the entrepreneur will already have background information that can support their decision making (Little, 2002). Urgency can be a justification for certain decisions, but it is not an excuse for not reviewing the data (Wallace, 2004). Indeed, even after a quick decision has been made; it is still possible to revisit the data to understand whether that was the right decision or not (Sakuramoto, 2005). Although it may be too late for the hastily made decision, it can prevent future problems because the decision-maker will have learnt from their mistakes (Noughabi and Arghami, 2011). The data science modalities that are on the market today can also cope with fast decision making since they provide information in real-time upon request (Helmreich, 2000). All that an executive has to do is check their smartphone and they will have important data that can impact on their decision making (Holmes, 2005). It is even possible to have a dedicated analyst that is consulted when decisions are being made (Little, 2002). Of course, the sense of urgency must be properly communicated because some analytics will continue working slowly and deliberately without recognizing the costs to the company if they do not keep pace with what is happening (Menke et al., 2007).

When making data-driven decisions, executives sometimes complain that they are not provided with the kind of guidance that they hoped for (Engelberger, 1982). The reports are either too detailed or too obscure (Howells and Wood, 1993). Others are merely looking for quick summaries that give them quick answers to complex problems (Jibril and Abdullah, 2013). It is important for executives to recognize the fact that the data analysis tends to be guided by the research brief that is provided by the client (Howells and Wood, 1993). This research brief may be a broad document that is used in a generic way or it could be specific when a problem arises and the decision-maker wants all the possible answers (Min et al., 2008). The complexity of the problem dictates the pace of responding and the complexity of the answer (Noughabi and Arghami, 2011). Therefore, executives must calibrate their thinking in ways that align with the kinds of business problems that they are presenting to the analyst (Min et al., 2009). The analyst does not actually make the final decision (Min et al., 2008). They merely support the decision-making process by presenting pertinent data (Ulloth, 1992).

Nevertheless, data science reports must be presented in formats that are accessible and understandable to the decisions makers (Hilbert and Lopez, 2011). They are a persuasive decision-making tool and these reports cannot fully maintain that role if they are not clear or do not speak to the issues that are important to the decision-maker (Lewis, 1996). It is also important to avoid overwhelming the decision-maker with multiple reports that are sometimes self-contradicting (Lyytinen et al., 2016). The reality is that managers tend to switch off if they are given a mini data deluge by a data analyst that does not know how to curate information is that it is the most relevant to the situation (Lewis, 1996). That does not mean that the analyst is required to massage the data in order to tell the client what they want to hear (Zhang and Chen, 2015). Data science speaks for itself and does not need to be embellished in order to carry its message across (Wallace, 2004). In any case, many of the things that will be highlighted in the report are factually and cannot be changed by subjective negative feelings about them (Sakuramoto, 2005). Instead, the business should take corrective, preventative, and precaution actions against negative outcomes that are highlighted in the report (Mieczakowski et al., 2011).

8.4. MAINTAINING RECORDS OF DECISION RATIONALE

We have already hinted on the benefits of maintain records of the rationale that underpins decisions that are based on data (Engelberger, 1982). This is very important for those companies that hope to survive beyond the exit of the original executives (Hilbert and Lopez, 2011).

It should be an implicit goal of an organization to ensure that it can survive its founders (Holmes, 2005). Otherwise, there would be multiple businesses that open and close the moment that their creators are no longer engaged with them (Miller, 2014).

A durable business will be based on systems and procedures that can outlast changes in personnel (Ruben and Lievrouw, n.d.). Data science is part of the durability and it has been implicated in succession planning for some of the larger organizations (Gibson and Brown, 2009).

Maintaining records is also a form of accountability to the stakeholders in the business who may include the owners, creditors, employees, and customers (Miller, 2014).

This comes into play when there is a query about the decision which has been taken or when the decision has led to some unexpected negative consequences (Bansal, 2013).

A record shows that the people who took the decision acted rationally and based on the information that was available to them at the time (Mieczakowski et al., 2011). Figure 8.4 demonstrates the logic of decision making from the broad to the narrow and specific. Keeping a record of all these transactions can play a role in helping to track decision and also support those who take over from the old guard (Miller, 2014).

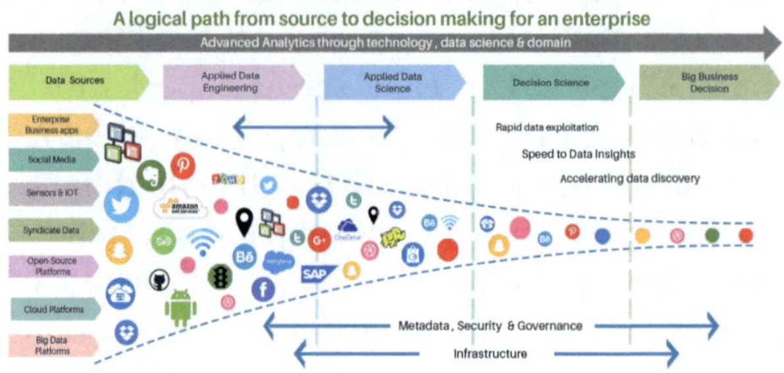

Figure 8.4. Logical pathways in decision making.

Source: Aptus Data Labs.

This is not about creating unnecessary layers of bureaucracy as would be the case in a public sector organization (Holmes, 2005). Rather it is about telling a story without gaps so that those who follow can continue that story (Jansen et al., 2008). It can also become an accountability measure for those that have a stake in the business to understand that the decisions which are consequential in that organization are never taken arbitrarily (Menke et al., 2007). The government as a regulator of the business environment may also be interested in understanding how decisions are made just in case there are questions of liability (Min et al., 2008). Good record-keeping might save the organization from fines or other legal penalties if they have taken a decision that is later found to be incompatible with the administrative regime in that locality (McFarlane, 2010). It is not just about keeping all the raw data for later reference. The analyst must chart their sources and the processes that they used in order to come to a final decision (Little, 2002). Besides, it might be useful to consider the alternatives so that the decision can be reconfigured if the feedback from the environment calls for such a response (Noughabi and Arghami, 2011). The record of data-driven decisions then ends up being a strong narrative about the operations of the company (Kees et al., 2015). This will give lenders, owners, and customer's confidence that the business is on the right footing and can survive in the future (Min et al., 2008).

8.5. LESS SUBJECTIVITY AND MORE OBJECTIVITY IN DECISION-MAKING

It is fitting that the conclusion to this book focuses on the core role of data science in decision making (Hilbert and Lopez, 2011). Good data science reduces the subjectivity that exists in business and instead promotes some level of objectivity (McFarlane, 2010). Perhaps this is an offshoot of the ontological and epistemological positioning of data science in which the methodologies emphasis an objective truth that can be gleaned through carefully organized research (McFarlane, 2010). Some might argue that subjectivity is inevitable in business since it is people that make the business (Stone et al., 2015). We know that human beings have their own biases and socialization which affect the way in which they view the world or respond to it (Cappellin and Wink, 2009). The imposition of strict objectivity on such a world might be counterproductive. On the other hand, we also know that data science is one of the many variables that contribute to the success of a business (Menke et al., 2007). Therefore, in rely on data science; decision-makers are engaging in a form of triangulation which moderates the biases of the human decision-makers with the objective data that is coming out of the environment (Mieczakowski et al., 2011). Figure 8.5 highlights the comparative differences between objective and subjective decision making.

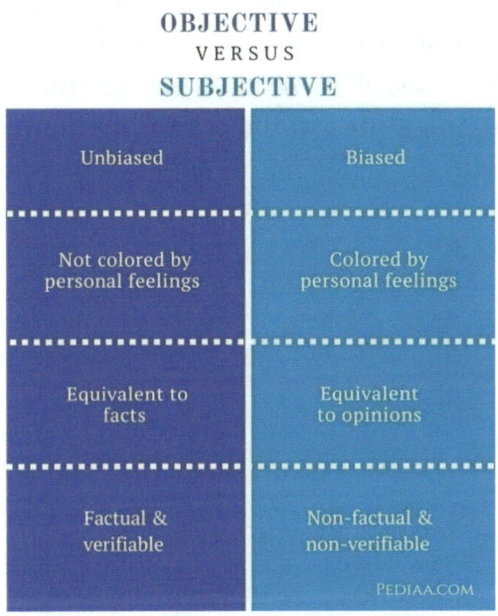

Figure 8.5. Objective and subjective decision making.

Source: Pediaa.

There are arguments to be had about whether or not objectivity is better than subjectivity in decision making (Cappellin and Wink, 2009). However, the evidence shows that data-driven organizations are able to survive in many business environments and that they do this by relying on information about those environments in order to make decisions (Howells and Wood,

1993). That has been the premise of this book and it should be the premise of all decision making in any given organization (Jibril and Abdullah, 2013). Intuition and other subjective measures are important, but they should not be the only consideration (Miller, 2014).

CHAPTER 8: SUMMARY

This final chapter aimed to demonstrate the different ways in which data science supports decision making. The chapter showed that data science helps to open up the perspectives of decision-makers so that they are able to consider all the available options before settling on the ones that seem to work. Moreover, data science provides a rationale and justifications for decisions which can help in succession planning and responding to queries from regulators. In order to achieve these benefits, the people that are doing the data analysis and decision making must ensure that they maintain good records of the decision-making process. Finally, this book does not make concrete conclusions about the superiority of objectivity to subjectivity in decision making. However, it advocates for complementarily in order to sustaining modern organizations in the era of big data.

CONCLUDING REMARKS

I hope you have enjoyed this book and that it will inspire you to research more about data science in general and its application to decision making in particular. This book was conceived from the standpoint that business and organizational management is a lot more than intuition and natural talent. We have an environment that is continuously producing information that can be useful for us when making a decision. It is a pity that consumers are using this information in order to make purchasing decisions, yet some businesses are reluctant to avail themselves of the opportunities that big data analysis gives in terms of improving their approach to decision making. This book is meant to inspire students, practitioners, entrepreneurs, and leaders to pay a lot more information to information in as far as, it pertains to the decisions that they make.

BIBLIOGRAPHY

1. Abu-Saifan, S., (2012). Social entrepreneurship: Definition and boundaries. *Technology Innovation Management Review,* pp. 22–27.
2. Awang, A. H., Hussain, M. Y., & Malek, J. A., (2013). Knowledge transfer and the role of local absorptive capability at science and technology parks. *The Learning Organization: An International Journal, 20*(4/5), 291–307.
3. Bachman, L. R., (2013). New professionalism: The post-industrial context. *Building Research and Information, 41*(6), 752–760.
4. Bansal, P., (2013). *Emerging Technology in Fashion Retail Business.* [Online] Available at: http://www.slideshare.net/bansalpan/emerging-technology-in-fashion-retail-business (Accessed on 18 December 2019).
5. Berker, T., Hartmann, M., Punie, Y., & Ward, K. J., (2006). *Domestication of Media and Technology.* Maidenhead: Open University Press.
6. Boase, J., (2008). Personal networks and the personal communication system. *Information, Communication and Society, 11*(4), pp. 490–508.
7. Cappellin, R., & Wink, R., (2009). *International Knowledge and*

Innovation Networks: Knowledge Creation and Innovation in Medium-Technology Clusters. Cheltenham: Edward Elgar.

8. Carlson, J. R., (1995). *Channel Expansion Theory: A Dynamic View of Media and Information Richness Perceptions.* Gainesville: Florida State University Press.
9. Carr, D., (2010). Time and technology: Addressing changing demands. In: Shea, C. M., & Garson, D. G., (eds.), *Handbook of Public Information Systems* (pp. 261–272). Boca Raton (FL): CRC Press.
10. Chesbrough, H. W., (2005). *Open Innovation: The New Imperative for Creating and Profiting from Technology* (1st Paper edn.). Boston (MA): Harvard Business Review Press.
11. Chiu, P. S., et al., (2016). Implementation and evaluation of mobile e-books in a cloud bookcase using the information system success model. *Library Hi Tech., 34*(2), pp. 207–223.
12. Davis, C., Xing, X., & Qian, Y., (2014). *Facebook and Corporate Social Responsibility: How should CSR be Enacted at Facebook with Regards to Customer Data?* Kindle ed. Online: Amazon Digital Services.
13. Dutse, A. Y., (2013). Linking absorptive capacity with innovative capabilities: A survey of manufacturing firms in Nigeria. *International Journal of Technology Management and Sustainable Development, 12*(2), 167–183.
14. Ellison, N. B., (2004). *Telework and Social Change: How Technology is Reshaping the Boundaries Between Home and Work.* Westpoint (CT): Praeger.
15. Engelberger, J. F., (1982). *Robotics in Practice: Future Capabilities,* s.l.: Electronic Servicing and Technology magazine.
16. Evans, R., (2009). *Balfour Beatty Among Firms That Bought Information on Workers.* London: The Guardian.
17. Gibson, W., & Brown, A., (2009). *Working With Qualitative Data.* London: SAGE.
18. Gilks, P., (2016). *Barclays Innovates with Customer Insight from Teradata and Tableau* [Interview] (2nd February 2016).
19. Hair, J. F., (2010). *Multivariate Data Analysis.* Upper Saddle River (NJ): Prentice Hall.
20. Hamari, J., Sjöklint, M., & Ukkonen, A., (2015). The sharing economy:

Why people participate in collaborative consumption. *Journal of the Association for Information Science and Technology,* Volume Forthcoming.

21. Helmreich, S., (2000). Flexible infections: Computer viruses, human bodies, nation-states, evolutionary capitalism. *Science, Technology, and Human Values, 25*(4), pp. 472–491.
22. Hilbert, M., & Lopez, P., (2011). The world's technological capacity to store, communicate, and compute information. *Science, 332*(6025), pp. 60–65.
23. Holmes, D., (2005). *Communication Theory: Media, Technology, and Society.* London: Sage Publications.
24. Howells, J., & Wood, M., (1993). *The Globalization of Production and Technology.* London: Belhaven Press.
25. Ifinedo, P., (2016). Applying uses and gratifications theory and social influence processes to understand students' pervasive adoption of social networking sites: Perspectives from the Americas. *International Journal of Information Management, 36*(2), pp. 192–194.
26. Jansen, B. J., Booth, D., & Spink, A., (2008). Determining the informational, navigational, and transactional intent of Web queries. *Information Processing and Management, 44*(3), pp. 1251–1266.
27. Jibril, T. A., & Abdullah, M. H., (2013). Relevance of emoticons in computer-mediated communication contexts: An overview. *Jibril, 9*(4).
28. Kees, A., Oberländer, A. M., Röglinger, M., & Rosemann, M., (2015). *Understanding the Internet of Things: A Conceptualization of Business-to-Thing (B2T) Interactions.* Münster (GER), Proceedings of the 23rd European Conference on Information Systems (ECIS).
29. Kim, S. Y., & Jeon, J. C., (2013). The effect of consumer tendency for Masstige brand on purchasing patterns-focusing on mediating effect of Massitige brand image. *Advances in Information Sciences and Service Sciences, 5*(15), pp. 343–355.
30. Kirchweger, S., Eder, M., & Kantelhardt, J., (2015). *Modeling the Effects of Low-Input Dairy Farming Using Bookkeeping Data from Austria Conference.* Milan, Italy, International Association of Agricultural Economist.
31. Kobie, N., (2015). *Get Yourself Connected: Is the Internet of Things the Future of Fashion?* [Online] Available at: http://www.theguardian.com/technology/2015/apr/21/internet-of-things-future-fashion

(Accessed on 18 December 2019).

32. Lewis, T., (1996). Studying the impact of technology on work and jobs. *Journal of Industrial Teacher Education, 33*(3), pp. 44–65.
33. Little, B., (2002). Harnessing learning technology to succeed in business. *Industrial and Commercial Training, 34*(2), pp. 76–80.
34. Lyytinen, K., Yoo, Y., & Boland Jr., R. J., (2016). Digital product innovation within four classes of innovation networks. *Information Systems Journal, 26*(1), pp. 47–75.
35. Malathy, S., & Kantha, P., (2013). Application of mobile technologies to libraries. *Journal of Library and Information Technology, 33*(5), p. DESIDOC.
36. McFarlane, D. A., (2010). Social communication in a technology-driven society: A philosophical exploration of factor-impacts and consequences. *American Communication Journal, 12*.
37. Menke, M., Xu, Q., & Gu, L., (2007). An analysis of the universality, flexibility, and agility of total innovation management: A case study of Hewlett–Packard. *Journal of Technology Transfer, 32*(1), pp. 49–62.
38. Mieczakowski, A., Goldhaber, T., & Clarkson, J., (2011). *Culture, Communication and Change: Summary of an Investigation of the Use and Impact of Modern Media and Technology in Our Lives.* Stoke-on-Trent: Engineering Design Centre.
39. Miller, J. B., (2014). *Internet Technologies and Information Services.* Santa Barbara (CA): ABC-CLIO.
40. Min, H., Min, H., Joo, S. J., & Kim, J., (2008). Data envelopment analysis for establishing the financial benchmark of Korean hotels. *International Journal of Services and Operations Management, 4*(2), p. 201.
41. Min, H., Min, H., Joo, S. J., & Kim, J., (2009). Evaluating the financial performances of Korean luxury hotels using data envelopment analysis. *The Service Industries Journal, 29*(6), 835–845.
42. Mosher, G. A., (2013). Trust, safety, and employee decision-making: A review of research and discussion of future directions. *The Journal of Technology, Management, and Applied Engineering, 29*(1).
43. Noughabi, H. A., & Arghami, N. R., (2011). Monte Carlo comparison of seven normality tests. *Journal of Statistical Computation and Simulation, 81*(8), 965–972.
44. Rachuri, K. K. et al., (2010). *Emotion Sense: A Mobile Phones Based*

Adaptive Platform for Experimental Social Psychology Research (pp. 281–290). Copenhagen, ACM.

45. Ruben, B. D., & Lievrouw, L. A., (n.d.). *Meditation, Information and Communication: Information and Behavior* (Vol. 3). London: Transaction Publishers.
46. Sakuramoto, N., (2005). Development of bookkeeping and management analysis system fou small-scall farmers (pocket bookkeeping system). *Agricultural Information Research (Japan).*
47. Schute, S., (2013). *Is Technology Moving Too Fast?* [Online] Available at: http://realbusiness.co.uk/article/22378-is-technology-moving-too-fast (Accessed on 18 December 2019)
48. Sin, S. C. J., (2016). Social media and problematic everyday life information□seeking outcomes: Differences across use frequency, gender, and problem□solving styles. *Journal of the Association for Information Science and Technology, 67*(8), 1793–1807.
49. Sinclaire, J., & Vogus, C., (2011). Adoption of social networking sites: An exploratory adaptive structuration perspective for global organizations. *Information Technology and Management, 12*(4), 293–314.
50. Sobh, R., & Perry, C., (2006). Research design and data analysis in realism research. *European Journal of Marketing, 40*(11/12), 1194–1209.
51. Spiekermann, S., Krasnova, H., Koroleva, K., & Hildebrand, T., (2010). Online social networks: Why we disclose. *Journal of Information Technology, 25*(2), 109–125.
52. Stone, D. L., Deadrick, D. L., Lukaszewskic, K. M., & Johnson, R., (2015). The influence of technology on the future of human resource management. *Human Resource Management: Past, Present and Future, 25*(2), 216–231.
53. Tarafdar, M., D'Arcy, J., Turel, O., & Gupta, A., (2014). The dark side of information technology. *MIT Sloan Management Review.* Research Feature (Winter).
54. Trottier, D., (2014). Crowdsourcing CCTV surveillance on the internet. *Information, Communication and Society, 17*(5), 609–626.
55. Ulloth, D. R., (1992). *Communication Technology: A Survey.* Lanham: University Press of America.
56. Van Deursen, C., & Van Dijk, (2014). Internet skills, sources of support

and benefiting from internet use. *International Journal of Human-Computer Interaction, 30*(4), pp. 278–290.

57. Van Nederpelt, P., & Daas, P., (2012). *49 Factors That Influence the Quality of Secondary Data Sources,* The Hague: Statistics Netherland.
58. Wallace, P., (2004). *The Internet in the Workplace: How New Technology is Transforming Work.* Cambridge: Cambridge University Press.
59. Zhang, Y., & Chen, J., (2015). Constructing scalable internet of things services based on their event-driven models. *Concurrency and Computation: Practice and Experience, 27*(17), 4819–4851.

INDEX

D

E

F

G

H

R

S

T

V

W

Printed in the United States
By Bookmasters